# 南洋花木

## 上海交通大学观赏树木图鉴
（中英双语）

ORNAMENTAL TREES OF SJTU

张荻 ◎ 编著

上海交通大学出版社
SHANGHAI JIAO TONG UNIVERSITY PRESS

## 内容提要

本书遵循科学、简洁、实用的原则进行编写,全书共收录73科188属314种观赏树木的信息,其中裸子植物采用郑万钧分类系统,被子植物采用恩格勒分类系统。全书采用彩色实拍图片展示树种的关键识别与观赏特征,并对各树种的主要形态特征、生态习性、园林应用与产地分布进行总结描述,将理论与实践教学相结合,以便于理解记忆、系统掌握。

本书可用作风景园林、观赏园艺、城市规划与环境设计等本科专业核心课程的配套补充教材,以加强理论与实践的结合,提升整体教学效果。同时,它也可作为相关专业人员与观赏植物爱好者的科普工具书。

## 图书在版编目(CIP)数据

南洋花木:上海交通大学观赏树木图鉴:汉、英 / 张荻编著. -- 上海:上海交通大学出版社,2024.6
ISBN 978-7-313-18602-7

Ⅰ.①南… Ⅱ.①张… Ⅲ.①上海交通大学-观赏树木-图集 Ⅳ.①S68-64

中国国家版本馆CIP数据核字(2023)第243725号

## 南洋花木——上海交通大学观赏树木图鉴(中英双语)
NANYANG HUAMU——SHANGHAI JIAOTONG DAXUE GUANSHANG SHUMU TUJIAN (ZHONGYINGSHUANGYU)

编　　著:张　荻

| | |
|---|---|
| 出版发行:上海交通大学出版社 | 地　　址:上海市番禺路951号 |
| 邮政编码:200030 | 电　　话:021-64071208 |
| 印　　制:上海锦佳印刷有限公司 | 经　　销:全国新华书店 |
| 开　　本:787mm×1092mm　1/16 | 印　　张:21.5 |
| 字　　数:505千字 | |
| 版　　次:2024年6月第1版 | 印　　次:2024年6月第1次印刷 |
| 书　　号:ISBN 978-7-313-18602-7 | |
| 定　　价:128.00元 | |

版权所有　侵权必究
告读者:如发现本书有印装质量问题请与印刷厂质量科联系
联系电话:021-56401314

# 前言

上海交通大学的前身南洋公学，创办于1896年，是一所历史悠久、享誉海内外的高等学府。历经120余年的奋发图强，上海交通大学已成为一所"综合性、研究型、国际化"的国内一流、国际知名大学。学校现有徐汇、闵行、黄浦、长宁、浦东等多个校区，总占地面积达300余万平方米。其中，闵行主校区自1985年起开始建设，如今已发展成为设施完善、环境优美的现代化大学校园。近年来，学校致力于建设绿色生态、富含人文底蕴的校园，并在教育部和农业农村部的支持下，实施了"园林植物标本园"工程。该工程共搜集了近500种观赏树木，分属81科262属，涵盖了上海市及周边地区常见的观赏树木种类，因此被誉为"上海市第三植物园"。这一园林空间逐渐形成了集教学、科研、科普、游憩功能于一体的开放式格局。园林植物标本园的建设堪称学校提升校园生态文化的一个缩影。

植物蕴藏着丰富的遗传信息和基因多样性，在生物圈、生态系统以及人居环境和社会经济发展中具有重要的基础支撑作用。十八大以来，我国积极倡导生态文明、美丽中国、大健康产业、两山理论以及"双碳"目标等国家发展战略，这些战略对国土空间和城乡景观在生态与康养方面的功能提出了更高的要求。植物、山水、建筑是构成人居环境景观的三大要素，其中植物是唯一具有生命特征的要素，在生态环境修复、景观营造、人民康养、社会可持续发展与满足人民文化需求方面发挥着不可或缺的作用。因此，对不同观赏植物的形态特征、观赏特性与生态习性等方面知识的掌握与学习，是营造可持续景观和进行生态设计的必要基础及有效途径。

观赏植物种类繁多，既涵盖系统性的理论知识，又具备很强的实践应用价值。在教学实践中，理论教学侧重于传授抽象的基础知识，而实践教学则着重于培养具体形态特征的观察能力和应用技能。本书收录了上海交通大学校园内73科188属314种观赏树木的实拍照片，详细展示了植株的整体形态、细部识别特征（包括干、枝、叶、花、果实等部分）以及它们在不同季节的观赏特点。此外，书中还补充了关于主要识别形态特征、生态习性、园林应用与产地分布的描述信息，力求内容准确、简洁易懂。此书可作为观赏植物学、园林树木学、植物生态设计、园林植物综合实践等课程的配套指导教材，旨在加强理论与实践教学之间的过渡与衔接，提升我校在人才培养中"知识探究"与"能力建设"的成效。

由于编者水平有限，疏漏与不当之处在所难免，敬请广大读者批评指正，不吝赐教。

张 荻<br>2024年6月于上海

# 上篇　裸子植物

## 1. 苏铁科

苏铁 *Cycas revoluta* ......2

## 2. 银杏科

银杏 *Ginkgo biloba* ......3

## 3. 松科

青杄 *Picea wilsonii* ......4
金钱松 *Pseudolarix amabilis* ......5
雪松 *Cedrus deodara* ......6
日本五针松 *Pinus parviflora* ......7
马尾松 *Pinus massoniana* ......8
黑松 *Pinus thunbergii* ......9
湿地松 *Pinus elliottii* ......10
白皮松 *Pinus bungeana* ......11

## 4. 杉科

杉木 *Cunninghamia lanceolata* ......12
日本柳杉 *Cryptomeria japonica* ......13
柳杉 *Cryptomeria japonica* var. *sinensis* ......14
落羽杉 *Taxodium distichum* ......15
墨西哥落羽杉 *Taxodium mucronatum* ......16
池杉 *Taxodium distichum* var. *imbricatum* ......17
水杉 *Metasequoia glyptostroboides* ......18
金叶水杉 *Metasequoia glyptostroboides* 'Gold Rush' ......19

台湾杉 *Taiwania cryptomerioides* .................................................. 20

### 5. 柏科

金塔侧柏 *Platycladus orientalis* 'Beverleyensis' .................................. 21
蓝冰柏 *Cupressus glabra* 'Blue Ice' ............................................. 22
圆柏 *Sabina chinensis* ......................................................... 23
龙柏 *Sabina chinensis* 'Kaizuka' ............................................... 24
金星球桧 *Sabina chinensis* 'Aureo-globosa' ..................................... 25
日本花柏 *Chamaecyparis pisifera* ............................................... 26
金线柏 *Chamaecyparis pisifera* 'Filifera Aurea' ................................ 27

### 6. 罗汉松科

罗汉松 *Podocarpus macrophyllus* ................................................ 28
竹柏 *Nageia nagi* .............................................................. 29

### 7. 三尖杉科

三尖杉 *Cephalotaxus fortunei* .................................................. 30

### 8. 红豆杉科

南方红豆杉 *Taxus wallichiana* var. *mairei* .................................... 31
东北红豆杉 *Taxus cuspidata* .................................................... 32
榧树 *Torreya grandis* .......................................................... 33

## 下篇 被子植物

### 9. 杨梅科

杨梅 *Myrica rubra* ............................................................. 36

### 10. 胡桃科

野核桃 *Juglans mandshurica* .................................................... 37

枫杨 *Pterocarya stenoptera* ............................................................................................... 38

## 11. 杨柳科

毛白杨 *Populus tomentosa* ............................................................................................. 39
中华红叶杨 *Populus deltoids* 'Zhonghua hongye' .......................................................... 40
加杨 *Populus × canadensis* ............................................................................................ 41
垂柳 *Salix babylonica* ..................................................................................................... 42
龙爪柳 *Salix matsudana* 'Tortuosa' ................................................................................ 43
银芽柳 *Salix × leucopithecia* .......................................................................................... 44
彩叶杞柳 *Salix integra* 'Hakuro Nishiki' ......................................................................... 45
红叶腺柳 *Salix chaenomeloides* 'Variegata' .................................................................... 46

## 12. 桦木科

桤木 *Alnus cremastogyne* ................................................................................................ 47
千金榆 *Carpinus cordata* ................................................................................................ 48

## 13. 壳斗科

板栗 *Castanea mollissima* ............................................................................................... 49
钩栗 *Castanopsis tibetana* .............................................................................................. 50
苦槠 *Castanopsis sclerophylla* ........................................................................................ 51
青冈栎 *Cyclobalanopsis glauca* ...................................................................................... 52
麻栎 *Quercus acutissima* ................................................................................................ 53
栓皮栎 *Quercus variabilis* .............................................................................................. 54
沼生栎 *Quercus palustris* ............................................................................................... 55
弗吉尼亚栎 *Quercus virginiana* ..................................................................................... 56

## 14. 榆科

榆树 *Ulmus pumila* ......................................................................................................... 57
榔榆 *Ulmus parvifolia* .................................................................................................... 58
榉树 *Zelkova serrata* ...................................................................................................... 59
珊瑚朴 *Celtis julianae* .................................................................................................... 60
小叶朴 *Celtis bungeana* ................................................................................................. 61

青檀 *Pteroceltis tatarinowii* .................................................................................. 62

### 15. 交让木科

交让木 *Daphniphyllum macropodum* ................................................................... 63

### 16. 杜仲科

杜仲 *Eucommia ulmoides* .................................................................................. 64

### 17. 桑科

桑树 *Morus alba* ............................................................................................... 65
构树 *Broussonetia papyrifera* ............................................................................ 66
无花果 *Ficus carica* ......................................................................................... 67
天仙果 *Ficus erecta* ......................................................................................... 68
薜荔 *Ficus pumila* ............................................................................................ 69
柘树 *Maclura tricuspidata* ................................................................................. 70

### 18. 木兰科

广玉兰 *Magnolia grandiflora* .............................................................................. 71
玉兰 *Magnolia denudata* ................................................................................... 72
紫玉兰 *Magnolia liliflora* ................................................................................... 73
二乔玉兰 *Magnolia soulangeana* ....................................................................... 74
白兰 *Michelia alba* ............................................................................................ 75
含笑 *Michelia figo* ............................................................................................ 76
金叶含笑 *Michelia foveolata* ............................................................................ 77
峨眉含笑 *Michelia wilsonii* ............................................................................... 78
鹅掌楸 *Liriodendron chinense* .......................................................................... 79
北美鹅掌楸 *Liriodendron tulipifera* ................................................................... 80

### 19. 蜡梅科

蜡梅 *Chimonanthus praecox* ............................................................................. 81
亮叶蜡梅 *Chimonanthus nitens* ......................................................................... 82

## 20. 樟科

樟树 Cinnamomum camphora ..................................................... 83
天竺桂 Cinnamomum japonicum ................................................ 84
浙江楠 Phoebe chekiangensis ..................................................... 85
红楠 Machilus thunbergii ............................................................. 86
薄叶润楠 Machilus leptophylla .................................................... 87
刨花楠 Machilus pauhoi ............................................................... 88
山胡椒 Lindera glauca ................................................................. 89
狭叶山胡椒 Lindera angustifolia ................................................. 90
月桂 Laurus nobilis ...................................................................... 91
舟山新木姜子 Neolitsea sericea ................................................... 92

## 21. 小檗科

豪猪刺 Berberis julianae ............................................................. 93
阔叶十大功劳 Mahonia bealei .................................................... 94
狭叶十大功劳 Mahonia fortunei ................................................. 95
南天竹 Nandina domestica .......................................................... 96

## 22. 猕猴桃科

中华猕猴桃 Actinidia chinensis ................................................... 97

## 23. 山茶科

山茶 Camellia japonica ............................................................... 98
油茶 Camellia oleifera ................................................................. 99
茶梅 Camellia sasanqua .............................................................. 100
杜鹃红山茶 Camellia azalea ....................................................... 101
厚皮香 Ternstroemia gymnanthera ............................................. 102

## 24. 藤黄科

金丝桃 Hypericum monogynum ................................................. 103
金丝梅 Hypericum patulum ........................................................ 104

## 25. 连香树科

连香树 Cercidiphyllum japonicum ..... 105

## 26. 悬铃木科

二球悬铃木 Platanus acerifolia ..... 106

## 27. 金缕梅科

枫香 Liquidambar formosana ..... 107
北美枫香 Liquidambar styraciflua ..... 108
细柄蕈树 Altingia gracilipes ..... 109
檵木 Loropetalum chinense ..... 110
红花檵木 Loropetalum chinense var. rubrum ..... 111
蜡瓣花 Corylopsis sinensis ..... 112
蚊母树 Distylium racemosum ..... 113
小叶蚊母树 Distylium buxifolium ..... 114

## 28. 虎耳草科

山梅花 Philadelphus incanus ..... 115
溲疏 Deutzia scabra ..... 116
雪球冰生溲疏 Deutzia gracilis 'Nikko' ..... 117
绣球花 Hydrangea macrophylla ..... 118
银边八仙花 Hydrangea macrophylla var. normalis 'Maculata' ..... 119

## 29. 海桐花科

海桐 Pittosporum tobira ..... 120
斑叶海桐 Pittosporum tobira 'Variegata' ..... 121

## 30. 蔷薇科

麻叶绣线菊 Spiraea cantoniensis ..... 122
粉花绣线菊 Spiraea japonica ..... 123

珍珠花 *Spiraea thunbergii* ...................................................................................... 124

风箱果 *Physocarpus amurensis* ............................................................................... 125

平枝栒子 *Cotoneaster horizontalis* ........................................................................... 126

火棘 *Pyracantha fortuneana* ..................................................................................... 127

小丑火棘 *Pyracantha fortuneana* 'Harlequin' ........................................................... 128

山楂 *Crataegus pinnatifida* ....................................................................................... 129

枇杷 *Eriobotrya japonica* .......................................................................................... 130

石楠 *Photinia serrulata* ............................................................................................. 131

红叶石楠 *Photinia* × *fraseri* ...................................................................................... 132

椤木石楠 *Photinia davidsoniae* ................................................................................. 133

日本贴梗海棠 *Chaenomeles japonica* ...................................................................... 134

木瓜 *Chaenomeles sinensis* ..................................................................................... 135

垂丝海棠 *Malus halliana* ........................................................................................... 136

玫瑰 *Rosa rugosa* ..................................................................................................... 137

月季花 *Rosa chinensis* ............................................................................................. 138

安吉拉月季 *Rosa hybrida* 'Angela' ........................................................................... 139

粉团蔷薇 *Rosa multiflora* var. *cathayensis* .............................................................. 140

七姊妹 *Rosa multiflora* var. *carnea* .......................................................................... 141

木香 *Rosa banksiae* .................................................................................................. 142

山木香 *Rosa cymosa* ................................................................................................ 143

棣棠 *Kerria japonica* ................................................................................................. 144

插田泡 *Rubus coreanus* ............................................................................................ 145

桃 *Amygdalus persica* ............................................................................................... 146

紫叶碧桃 *Amygdalus persica* 'Atropurpurea' ............................................................ 147

杏 *Armeniaca vulgaris* .............................................................................................. 148

宫粉梅 *Armeniaca mume* var. *alphandii* .................................................................. 149

李 *Prunus salicina* .................................................................................................... 150

紫叶李 *Prunus cerasifera* f. *atropurpurea* ............................................................... 151

杜梨 *Pyrus betulaefolia* ............................................................................................. 152

厚叶石斑木 *Rhaphiolepis umbellata* ......................................................................... 153

樱花 *Cerasus serrulata* ............................................................................................. 154

郁李 *Cerasus japonica* ............................................................................................. 155

## 31. 豆科

合欢 *Albizia julibrissin* ............................................................................................... 156

银荆 *Acacia dealbata* .................................................. 157

云实 *Caesalpinia decapetala* ........................................ 158

紫荆 *Cercis chinensis* ................................................. 159

紫叶加拿大紫荆 *Cercis canadensis* 'Purpurea' ................ 160

皂荚 *Gleditsia sinensis* ............................................... 161

山皂荚 *Gleditsia japonica* ........................................... 162

美国金叶皂荚 *Gleditsia triacanthos* 'Sunburst' ............... 163

龙牙花 *Erythrina corallodendron* ................................ 164

紫穗槐 *Amorpha fruticosa* ......................................... 165

紫藤 *Wisteria sinensis* ............................................... 166

常春油麻藤 *Mucuna sempervirens* .............................. 167

龙爪槐 *Sophora japonica* f. *Pendula* .......................... 168

蝴蝶槐 *Sophora japonica* var. *oligophylla* ................... 169

刺槐 *Robinia pseudoacacia* ....................................... 170

锦鸡儿 *Caragana sinica* ............................................ 171

红豆树 *Ormosia hosiei* .............................................. 172

胡枝子 *Lespedeza bicolor* ......................................... 173

花木蓝 *Indigofera kirilowii* ........................................ 174

### 32. 大戟科

山麻杆 *Alchornea davidii* .......................................... 175

乌桕 *Sapium sebiferum* ............................................ 176

重阳木 *Bischofia polycarpa* ....................................... 177

### 33. 芸香科

野花椒 *Zanthoxylum simulans* ................................... 178

竹叶椒 *Zanthoxylum armatum* ................................... 179

枳 *Poncirus trifoliata* ................................................ 180

柑橘 *Citrus reticulata* ............................................... 181

酸橙 *Citrus aurantium* .............................................. 182

甜橙 *Citrus sinensis* ................................................. 183

### 34. 苦木科

臭椿 *Ailanthus altissima* ............................................ 184

## 35. 楝科

苦楝 *Melia azedarach* ........................................................................................................ 185

## 36. 漆树科

黄连木 *Pistacia chinensis* ................................................................................................. 186
黄栌 *Cotinus coggygria* .................................................................................................... 187
南酸枣 *Choerospondias axillaris* ..................................................................................... 188
元宝枫 *Acer truncatum* ..................................................................................................... 189
秀丽槭 *Acer elegantulum* .................................................................................................. 190
三角枫 *Acer buergerianum* ............................................................................................... 191
鸡爪槭 *Acer palmatum* ...................................................................................................... 192
红枫 *Acer palmatum* 'Atropurpureum' ............................................................................ 193
羽毛枫 *Acer palmatum* 'Dissectum' ................................................................................. 194
樟叶槭 *Acer cinnamomifolium* ......................................................................................... 195
茶条槭 *Acer ginnala* .......................................................................................................... 196
苦茶槭 *Acer ginnala* ssp. *theiferum* ............................................................................... 197
羽叶槭 *Acer negundo* ........................................................................................................ 198
罗浮槭 *Acer fabri* ............................................................................................................... 199

## 37. 无患子科

复羽叶栾树 *Koelreuteria bipinnata* ................................................................................. 200
无患子 *Sapindus mukorossi* .............................................................................................. 201

## 38. 七叶树科

七叶树 *Aesculus chinensis* ................................................................................................ 202

## 39. 冬青科

枸骨 *Ilex cornuta* ................................................................................................................ 203
无刺枸骨 *Ilex cornuta* var. *fortunei* ................................................................................ 204
大叶冬青 *Ilex latifolia* ....................................................................................................... 205
龟甲冬青 *Ilex crenata* var. *convexa* ................................................................................ 206

## 40. 卫矛科

大花卫矛 *Euonymus grandiflorus* .................................................. 207
卫矛 *Euonymus alatus* .................................................. 208
丝棉木 *Euonymus maackii* .................................................. 209
扶芳藤 *Euonymus fortunei* .................................................. 210
大叶黄杨 *Euonymus japonicus* .................................................. 211

## 41. 黄杨科

小叶黄杨 *Buxus microphylla* .................................................. 212

## 42. 鼠李科

枳椇 *Hovenia acerba* .................................................. 213
枣树 *Ziziphus jujuba* .................................................. 214
马甲子 *Paliurus ramosissimus* .................................................. 215
圆叶鼠李 *Rhamnus globosa* .................................................. 216
长叶冻绿 *Rhamnus crenata* .................................................. 217

## 43. 葡萄科

葡萄 *Vitis vinifera* .................................................. 218

## 44. 杜英科

尖叶杜英 *Elaeocarpus apiculatus* .................................................. 219
中华杜英 *Elaeocarpus chinensis* .................................................. 220

## 45. 锦葵科

木芙蓉 *Hibiscus mutabilis* .................................................. 221
木槿 *Hibiscus syriacus* .................................................. 222
海滨木槿 *Hibiscus hamabo* .................................................. 223

## 46. 椴树科

心叶椴 *Tilia cordata* ...... 224

## 47. 梧桐科

梧桐 *Firmiana platanifolia* ...... 225
梭罗树 *Reevesia pubescens* ...... 226

## 48. 瑞香科

结香 *Edgeworthia chrysantha* ...... 227

## 49. 胡颓子科

胡颓子 *Elaeagnus pungens* ...... 228
金边胡颓子 *Elaeagnus pungens* 'Goldrim' ...... 229
木半夏 *Elaeagnus multiflora* ...... 230

## 50. 柽柳科

柽柳 *Tamarix chinensis* ...... 231

## 51. 千屈菜科

紫薇 *Lagerstroemia indica* ...... 232
福建紫薇 *Lagerstroemia limii* ...... 233

## 52. 桃金娘科

大叶桉 *Eucalyptus robusta* ...... 234
红千层 *Callistemon rigidus* ...... 235
轮叶赤楠 *Syzygium buxifolium* var. *verticillatum* ...... 236
香桃木 *Myrtus communis* ...... 237
花叶香桃木 *Myrtus communis* 'Variegata' ...... 238
小叶香桃木 *Myrtus communis* var. *microphylla* ...... 239

菲油果 *Feijoa sellowiana* ........................................ 240

松红梅 *Leptospermum scoparium* .......................... 241

### 53. 石榴科

石榴 *Punica granatum* ............................................ 242

### 54. 蓝果树科

喜树 *Camptotheca acuminata* ................................. 243

### 55. 山茱萸科

洒金东瀛珊瑚 *Aucuba japonica* var. *variegata* ....... 244

红瑞木 *Cornus alba* ................................................ 245

光皮梾木 *Cornus wilsoniana* .................................. 246

香港四照花 *Dendrobenthamia hongkongensis* ....... 247

山茱萸 *Macrocarpium officinale* ............................. 248

### 56. 五加科

八角金盘 *Fatsia japonica* ........................................ 249

熊掌木 *Fatshedera lizei* .......................................... 250

### 57. 八角枫科

八角枫 *Alangium chinense* ..................................... 251

### 58. 大风子科

山桐子 *Idesia polycarpa* ......................................... 252

### 59. 清风藤科

泡花树 *Meliosma cuneifolia* ................................... 253

红枝柴 *Meliosma oldhamii* .................................... 254

## 60. 唇形科

水果蓝 *Teucrium fruticans* .......... 255

## 61. 杜鹃花科

杜鹃花 *Rhododendron simsii* .......... 256
毛白杜鹃 *Rhododendron mucronatum* .......... 257
羊踯躅 *Rhododendron molle* .......... 258
马醉木 *Pieris japonica* .......... 259

## 62. 紫金牛科

朱砂根 *Ardisia crenata* .......... 260

## 63. 柿树科

柿 *Diospyros kaki* .......... 261
君迁子 *Diospyros lotus* .......... 262
油柿 *Diospyros oleifera* .......... 263
老鸦柿 *Diospyros rhombifolia* .......... 264

## 64. 安息香科

秤锤树 *Sinojackia xylocarpa* .......... 265

## 65. 木犀科

美国白蜡树 *Fraxinus americana* .......... 266
金钟花 *Forsythia viridissima* .......... 267
雪柳 *Fontanesia fortunei* .......... 268
桂花 *Osmanthus fragrans* .......... 269
女贞 *Ligustrum lucidum* .......... 270
金森女贞 *Ligustrum japonicum* 'Howardii' .......... 271
金叶女贞 *Ligustrum* × *vicaryi* .......... 272

红药小蜡 *Ligustrum sinense* 'Multiflorum' .................................................. 273
云南黄馨 *Jasminum mesnyi* .................................................. 274
浓香茉莉 *Jasminum odoratissimum* .................................................. 275
白丁香 *Syringa oblata* 'Alba' .................................................. 276

## 66. 马钱科

大叶醉鱼草 *Buddleja davidii* .................................................. 277

## 67. 夹竹桃科

络石 *Trachelospermum jasminoides* .................................................. 278
花叶蔓长春 *Vinca major* 'Variegata' .................................................. 279
夹竹桃 *Nerium oleander* .................................................. 280

## 68. 茜草科

水团花 *Adina pilulifera* .................................................. 281
六月雪 *Serissa japonica* .................................................. 282
栀子 *Gardenia jasminoides* .................................................. 283
水栀子 *Gardenia jasminoides* var. *radicans* .................................................. 284

## 69. 马鞭草科

马缨丹 *Lantana camara* .................................................. 285
华紫珠 *Callicarpa cathayana* .................................................. 286
杜虹花 *Callicarpa formosana* .................................................. 287
海州常山 *Clerodendrum trichotomum* .................................................. 288
大青 *Clerodendrum cyrtophyllum* .................................................. 289
黄荆 *Vitex negundo* .................................................. 290
牡荆 *Vitex negundo* var. *cannabifolia* .................................................. 291

## 70. 泡桐科

毛泡桐 *Paulownia tomentosa* .................................................. 292

## 71. 紫葳科

梓树 *Catalpa ovata* ......293
楸树 *Catalpa bungei* ......294
黄金树 *Catalpa speciosa* ......295
凌霄 *Campsis grandiflora* ......296

## 72. 忍冬科

红王子锦带 *Weigela florida* 'Red Prince' ......297
花叶锦带 *Weigela florida* 'Variegata' ......298
海仙花 *Weigela coraeensis* ......299
猬实 *Kolkwitzia amabilis* ......300
大花六道木 *Abelia* × *grandiflora* ......301
金银花 *Lonicera japonica* ......302
金银忍冬 *Lonicera maackii* ......303
郁香忍冬 *Lonicera fragrantissima* ......304
金红久忍冬 *Lonicera* × *heckrottii* ......305
接骨木 *Sambucus williamsii* ......306
日本珊瑚树 *Viburnum odoratissimum* var. *awabuki* ......307
木本绣球 *Viburnum macrocephalum* ......308
琼花 *Viburnum macrocephalum* f. *keteleeri* ......309
天目琼花 *Viburnum sargentii* ......310
枇杷叶荚蒾 *Viburnum rhytidophyllum* ......311
地中海荚蒾 *Viburnum tinus* ......312
蝴蝶绣球 *Viburnum plicatum* ......313
蝴蝶戏珠花 *Viburnum plicatum* f. *tomentosum* ......314
七子花 *Heptacodium miconioides* ......315

## 73. 棕榈科

棕榈 *Trachycarpus fortunei* ......316
加那利海枣 *Phoenix canariensis* ......317

**参考文献** ......318
**观赏树木中文名索引** ......319

上篇

裸子植物

# 1. 苏铁科

## 苏铁 | *Cycas revoluta*

**别名**：凤尾蕉、凤尾松、铁树　　　　　　**科属**：苏铁科苏铁属

**形态特征**：常绿，茎干呈圆柱状。羽状叶从茎的顶部生出，叶长 0.7～1.8 m，呈 "V" 字形伸展；小叶 40～100 对以上，厚革质，线形，坚硬而有光泽，初生时边缘显著反卷。雄球花圆柱形，黄色，直立于茎顶；雌球花扁球形，大孢子叶宽卵形，密被黄褐色绒毛。

**习性、应用及产地分布**：阳性树种，生长缓慢，材质密度大，如铁入水即沉，因此得名"铁树"。它喜铁质酸性土壤，但在长江以北地区由于受积温影响，难以开花，当温度降至 0℃时，叶片会受害。国家一级保护植物。产于中国福建沿海低山地区，长江流域以北地区常盆栽观赏，温室越冬。

**Description:** Evergreen, trunk cylindric. Leaves 40–100 or more, 1-pinnate, 0.7–1.8 m long, strongly "V"-shaped in cross section, leathery, linear, hard, semiglossy, margin strongly recurved when young. Microsporophylls ovoid-cylindric, yellow, erect at stem apex, ellipsoid, somewhat compressed. Megasporophylls broadly ovoid, yellow to pale brown, densely tomentose.

# 2. 银杏科

## 银杏 | *Ginkgo biloba*

**别名：** 白果、公孙树　　　　　　**科属：** 银杏科银杏属

**形态特征：** 落叶乔木，雌雄异株，稀同株。树干端直，具长枝、短枝之分。叶扇形，先端常2裂，于长枝上螺旋状散生，在短枝上3～5枚成簇生状，叶脉叉状并列；叶柄细长。种子核果状，外种皮肉质，被白粉；中种皮骨质，白色；内种皮膜质，淡红褐色。花期4—5月，种子9—10月成熟。

**习性、应用及产地分布：** 温带及亚热带阳性树种。喜光，耐干旱，不耐积水，深根性，耐移植。抗燃性强，少病虫危害。是著名的行道树种，寿命长达千年；叶形秀美，秋叶金黄，适宜作庭荫树、行道树或独赏树。沈阳以南地区可栽培。最早出现于3.45亿年前，有"活化石"之称。中国特有的单种属，国家一级重点保护植物。

**Description:** Known as the "living fossil". Trees, deciduous, bark light gray or grayish brown, longitudinally fissured especially on old trees; crown conical initially, finally broadly ovoid. Leaves scalloped, apex 2 splits, spirally scattered on the long branches, 3–5 clusters on the short branches; petiole slender. Drupe-like seeds, sarcotesta yellow, sclerotesta white, endotesta pale reddish brown. Fl. (flowering) Apr–May, fr. (fruiting) Sep–Oct.

## 3. 松科

### 青杆 | *Picea wilsonii*

**别名**：华北云杉、细叶云杉　　　　　**科属**：松科云杉属

**形态特征**：常绿乔木，高达50 m。树皮灰色或暗灰色，裂成不规则鳞片块状脱落。1年生小枝淡黄灰色，近无毛，小枝上具显著的叶枕，顶端凸起。叶四棱状条形，长8～13（18）mm，宽1～2 mm，无白粉，螺旋状着生于叶枕之上。球果圆柱状长卵圆形，长5～8 cm，径2.5～4 cm，成熟前绿色，熟时黄褐色，10月成熟。

**习性、应用及产地分布**：阳性树种。性强健，喜凉爽、湿润气候，耐寒。树形整齐，枝叶细密，蓝灰色针叶更具观赏性，为常见的庭园绿化树种，可孤植、列植、对植。中国特有种，产于东北、华北、西北及四川东北部。

**Description:** Trees, evergreen, up to 50 m tall. Bark gray or dark gray, irregularly flaking, crown pyramidal; branchlets pale yellowish gray, glabrous, branches with prominent pulvinus convex. Leaves broadly quadrangular in cross section, 8–13 mm × 1–2 mm, spirally born above the pulvirus. Seed cones ovoid-oblong, 5–8 cm × 2.5–4 cm, seeds obovoid, 3–4 mm. Pollination Apr, seed maturity Oct.

## 金钱松 | *Pseudolarix amabilis*

**别名**：金松、水树  **科属**：松科金钱松属

**形态特征**：落叶大乔木，高达40 m。树冠宽塔形；树皮灰褐色，裂成不规则的鳞片状；枝不规则轮生。叶条形、柔软，簇生于短枝。秋叶金黄，犹如金钱落地，故名。球果卵圆形或倒卵形，熟时红褐色。花期4—5月，球果10月成熟。

**习性、应用及产地分布**：喜光，略耐低温，不耐干旱贫瘠；深根性，不耐移栽。产于长江中下游以南地区，浙江西天目山有天然林分布。世界五大公园树种之一。中国特有单种属。国家二级重点保护植物。

**Description:** Trees, deciduous, up to 40 m tall. Crown broadly conical; bark gray-brown, rough, scaly, flaking; branches irregularly in whorls. Leaves narrow, linear, soft, clustered on short branches. Seed cones green or yellow-green, maturing reddish brown, obovoid or ovoid. Pollination Apr–May, seed maturity Oct.

# 雪松 | *Cedrus deodara*

**别名：** 喜马拉雅雪松、喜马拉雅杉　　**科属：** 松科雪松属

**形态特征：** 常绿大乔木，高可达 60 m。树冠圆锥形；树皮灰褐色，裂成不规则的鳞状块片；具长、短枝，枝条基部芽鳞宿存，大枝平展或微下垂。叶针形，坚硬，在长枝上螺旋状排列，在短枝上簇生。雄球花和雌球花分别单生于短枝顶端，直立。球果宽椭圆形，长 7～12 cm，熟时红褐色。花期 10—11 月，球果次年 10 月成熟。

**习性、应用及产地分布：** 喜光，有一定耐寒力，不耐湿热。较耐旱，忌积水；对土壤要求不严，浅根性，抗风力弱，可监测大气二氧化硫（$SO_2$）污染。寿命可达 800 年，产于喜马拉雅山区，中国北京以南至长江中下游城市普遍栽植。世界五大公园树种之一，有"风景树皇后"之美誉。

**Description:** Large trees, evergreen, up to 60 m tall. Crown conical; bark dark gray, cracking into irregular scales; possess long and short branchlets, branches horizontal, slightly tilted or slightly pendulous. Leaves acicular, hard, radially spreading on long branchlets, in apparent fascicles on short branchlets. Seed cones shortly pedunculate, ovoid or broadly ellipsoid, 7–12 cm long, becoming reddish brown when ripe. Pollination Oct–Nov, seed maturity Oct of following year.

# 日本五针松 | *Pinus parviflora*

**别名：** 日本五须松、五钗松　　　　**科属：** 松科松属

**形态特征：** 常绿大乔木，高达25 m。树冠圆锥形；树皮暗灰色，裂成鳞状块片脱落；1年生枝绿色，后呈黄褐色，密被淡黄色柔毛。针叶5针1束，长3.5～5.5 cm，具3～6条灰白色气孔线，叶鞘早落。球果卵圆形，长4.0～7.5 cm，幼时紫褐色，成熟时黑褐色，开裂。花期5月，球果翌年6月成熟。

**习性、应用及产地分布：** 喜光，在阴湿之处生长不良；喜温暖、湿润气候。对土壤要求不严，生长缓慢，可制作盆景观赏。常用的观赏树种，适于小型庭院和山石配景。原产于日本，中国引种历史悠久。

**Description:** Large trees, evergreen, up to 25 m tall. Crown conical; bark dull gray, furrowed into scaly plates; 1st-year branchlets initially green, aging yellow-brown, densely pale-yellow pubescent. Needles 5 per bundle, 3.5–5.5 cm long, with 3–6 stomatal lines, base with sheath shed. Seed cones ovoid, 4.0–7.5 cm long, puce when young, black-brown at maturity, cracked. Pollination May, seed maturity Jun of following year.

# 马尾松 | *Pinus massoniana*

**别名**：青松、枞松　　　　　　　　**科属**：松科松属

**形态特征**：常绿大乔木，高达40 m。树皮红褐色，裂成不规则鳞片；大枝斜展或平展，1年生小枝轮生，淡黄褐色，无白粉。针叶2针1束，长12～20 cm，细柔，微下垂，两面具气孔线；叶鞘宿存。球果卵圆形或圆锥状卵形，长4～7 cm，具短柄，熟时栗褐色。花期4—5月，球果翌年10—12月成熟。

**习性、应用及产地分布**：亚热带强阳性树种。喜温暖、多雨气候，耐寒性较差。对土壤要求不严，耐干旱、瘠薄。深根性，主根、侧根均发达。树形高大雄伟，生长快、适应性强。中国特有种，是江南及华南自然风景区重要的风景林树种和荒山造林树种，适宜群植或林植于道旁、溪畔和谷中。

**Description:** Large trees, evergreen, up to 40 m tall. Bark red-brown, irregularly scaly and flaking; branches slightly tilted or horizontal,1st-year branchlets whorled, pale yellowish brown, not covered in white powder. Needles 2 per bundle, 12−20 cm long, soft, slightly pendulous, stomatal lines present on all surfaces, base with persistent sheath. Seed cones ovoid or conical-ovoid, 4−7 cm long, shortly pedunculate, turning chestnut brown at maturity. Pollination Apr−May, seed maturity Oct−Dec of following year.

# 黑松 | *Pinus thunbergii*

**别名：** 白芽松、日本黑松　　　　**科属：** 松科松属

**形态特征：** 常绿大乔木，高达30 m。幼树树皮暗灰色，老则灰黑色，呈块片状脱落；1年生枝淡褐黄色，无毛；冬芽银白色。针叶2针1束，长6～12 cm，粗硬。球果圆锥状卵圆形，长4～6 cm，具短柄；鳞脐微凹，具短刺，鳞盾微肥厚。花期4—5月，球果翌年10月成熟。

**习性、应用及产地分布：** 阳性树种。喜温暖湿润的海洋性气候。耐盐碱，能生长于海滨潮水侵袭之地，抗病虫害能力较强，是著名海岸防护林树种。原产于日本及朝鲜南部沿海地区，中国华东沿海各省均有引种栽培。

**Description:** Large trees, evergreen, up to 30 m tall. Bark dull gray when young, aging gray-black, scaly and deciduous; 1st-year branchlets pale brown-yellow, glabrous; winter buds silvery white. Needles 2 per bundle, 6–12 cm long, rigid. Seed cones conical-ovoid, 4–6 cm long, shortly peduncular; umbo slightly concave, spinescent. Pollination Apr–May, seed maturity Oct of following year.

# 湿地松 | *Pinus elliottii*

**别名：** 古巴松　　　　　　　**科属：** 松科松属

**形态特征：** 常绿大乔木，在原产地可高达 30 m。树皮灰褐色或暗红褐色，纵裂成鳞块状剥落；枝每年可生长 3～4 轮，小枝粗壮。针叶 2 针 1 束、3 针 1 束并存，长 18～30 cm，深绿色，刚硬，有气孔线，叶鞘长约 1.2 cm。球果圆锥状卵形，长 6.5～13 cm，具梗。花期 2—3 月，球果翌年 9 月成熟。

**习性、应用及产地分布：** 强阳性树种。速生，尤耐水湿，但长期积水则生长不良。主侧根较发达，抗风力较强。原产于北美亚热带低海拔的潮湿地带，中国长江流域至华南地区引种栽培，是中国引种最成功的树种之一。

**Description:** Large trees, evergreen, up to 30 m tall. Bark gray- or red-brown, furrowed into scaly plates and deciduous; branchlets producing 3 or 4 whorls each year, stout. Needles 2 or 3 per bundle, 18–30 cm long, dark green, rigid, stomatal lines present on all surfaces, base with persistent sheath 1.2 cm in length. Seed cones conical-ovoid, 6.5–13 cm long, pedunculate. Pollination Feb–Mar, seed maturity Sep of following year.

# 白皮松 | *Pinus bungeana*

**别名**：三针松、虎皮松、白骨松　　　　**科属**：松科松属

**形态特征**：常绿大乔木，高达30 m。树冠塔形至伞形；树皮不规则薄块状脱落，内皮淡黄绿色；老则树皮呈淡褐色或灰白色，白褐相间呈斑鳞状；幼树及1年生枝条灰绿色，光滑。针叶3针1束，粗硬，长5～10 cm，叶鞘早落。球果卵圆形或圆锥状卵形，长5～7 cm，熟时淡黄褐色，近无柄。花期4—5月，球果翌年10—11月成熟。

**习性、应用及产地分布**：喜光，适生于干冷气候。对土壤要求不严。深根性，抗风力较强，对二氧化硫污染及烟尘污染有较强抗性。树皮斑驳，与长白松、樟子松、赤松、欧洲赤松一起被称为"五大"美人松。多与假山、岩石相配，植于庭院、寺院以及园林之中。中国特有种，辽宁南部至长江流域广为栽植。

**Description:** Large trees, evergreen, up to 30 m tall. Crown pyramidal or umbrellalike; bark exfoliating in irregular, thin, scaly patches, inner bark pale yellow-green; aging pale brown or grayish white, saplings and 1st-year branchlets gray-green, glabrous. Needles 3 per bundle, stiff, 5–10 cm long, base with sheath shed. Seed cones ovoid or conical-ovoid, 5–7 cm long, pale yellowish brown at maturity, subsessile. Pollination Apr–May, seed maturity Oct–Nov of following year.

## 4. 杉科

### 杉木 | *Cunninghamia lanceolata*

**别名**：沙木、沙树、正木    **科属**：杉科杉木属

**形态特征**：常绿乔木，高达30 m。主干通直；树皮褐色，长条片状脱落；大枝平展轮生。叶线状披针形，长2～6 cm，先端刺状。雌球花单生或2～3朵簇生于枝顶。球果长2.5～5 cm，宿存；种子褐色。花期4月，球果10月下旬成熟。

**习性、应用及产地分布**：亚热带树种。较喜光，适宜肥沃疏松、排水良好的土壤。浅根性，侧根发达；速生，南方重要的用材林树种。中国秦岭、淮河以南有分布，英国、美国、日本等地有引种。

**Description:** Trees, evergreen, up to 30 m tall. Crown pyramidal, trunk straight, bark brown, longitudinally fissured, cracking into irregular flakes; branches whorled, horizontally spreading. Leaves narrowly linear-lanceolate, 2–6 cm long, apex usually spinescent. Pistillate cones fascicles terminal, singly or 2–3 together. Seed cones 2.5–5 cm long, persistent; seeds dark brown. Pollination Apr, seed maturity Oct.

## 日本柳杉 | *Cryptomeria japonica*

**别名：**孔雀松  **科属：**杉科柳杉属

**形态特征：**常绿大乔木，高达40 m。树冠尖塔形；树皮红褐色，裂成条片状脱落；大枝常轮状着生，水平开展或微下垂；一年生枝绿色。叶钻形，长0.4～2 cm，直伸，先端不内曲或微内曲。雄球花长椭圆形或圆柱形，雌球花圆球形。1年生球果黄绿色，成熟时深褐色，径1～3 cm；种鳞20～30个，成熟时先端向外反曲，每个种鳞有种子2～5粒。花期4月，球果10月成熟。

**习性、应用及产地分布：**喜温暖、湿润气候，适生于排水良好的砂质壤土，积水时易烂根。对二氧化硫等有毒气体的吸收能力强。是日本主要造林树种。树形高大、树姿秀丽，适于群植供荫庇及防风之用，庙宇及神社内多参天巨木；可净化空气，一些矮小的灌木品种可植于庭院、园路交叉口、道路两侧、公园、草坪边缘。原产于日本，欧洲、北美、尼泊尔、印度及中国东部地区一些城市多作观赏树栽培。

**Description:** Large trees, evergreen, up to 40 m tall. Crown pyramidal; bark reddish brown, fibrous, peeling off in strips; branches whorled, horizontally spreading or slightly pendulous, 1st-year branchlets green. Leaves subulate, 0.4–2 cm long, straight or slightly incurved. Male cones ovoid or ovoid-cylindric, female cones globose. 1st-year seed cones yellowish green, dark brown at maturity, 1–3 cm in diam. (diameter); cone scales 20–30, apex usually recurved at maturity. Seeds 2–5 per cone scale. Pollination Apr, seed maturity Oct.

# 柳杉 | *Cryptomeria japonica* var. *sinensis*

**科属：** 杉科柳杉属

**形态特征：** 常绿乔木，高可达50 m。树皮棕褐色，条状纵裂；小枝细长，明显下垂。叶线状锥形，长1～1.5 cm，先端内曲，有气孔线。是日本柳杉的一个变种。与原种的区别是，叶先端内曲。球果小，种鳞20个左右，每个种鳞具种子2粒。花期4月，果期10—11月。

**习性、应用及产地分布：** 喜欢温暖湿润气候及肥厚、排水良好的酸性土壤，特别适生于空气湿度大、夏季凉爽的山地环境，不耐寒，浅根系，生长较快，对二氧化硫抗性较强。树形高大、树姿秀丽，可植于庭院、路口、道路两侧、公园、草坪边缘。产于中国浙江西天目山、江西庐山、华南、华东、西南及陕西关中、河南郑州等地。

**Description:** Trees, evergreen. Bark reddish brown, fibrous, peeling off in strips; branchlets usually pendulous. Leaves subulate to linear, 1–1.5 cm long, apex incurved, stomatal bands with 2–8 rows of stomata on each surface. Seed cones, globose or subglobose, cone scales ca. 20, each bearing 2 seeds. Pollination Apr, seed maturity Oct–Nov.

# 落羽杉 | *Taxodium distichum*

**别名**：落羽松、美国水松　　　　　　**科属**：杉科落羽杉属

**形态特征**：落叶大乔木，高达50 m。树干基部膨大，常具屈膝状呼吸根；树皮棕色，长片状脱落；1年生小枝褐色，侧生小枝冬季脱落。叶条形、扁平，长1～1.5 cm，基部扭曲在小枝上排成2列，羽毛状，秋叶古铜色。球果圆形或卵圆形，径约2.5 cm，下垂，成熟后淡褐黄色，被白粉。花期4月下旬，球果10月成熟。

**习性、应用及产地分布**：强阳性树种。极耐水湿，抗风力强，速生。树形高耸挺秀，羽状叶轻柔；适于公园、河畔、湿地配置，且具防风、固堤功效。原产于美国东南部，中国长江流域及华南地区园林常见栽培，较池杉分布更广。

**Description:** Large trees, deciduous, up to 50 m tall. Crown broadly conical; trunk swollen and buttressed at base, pneumatophores present or absent around trunk; bark brown, peeling off in long strips; 1st-year branchlets brown, lateral branchlets deciduous in winter. Leaves linear or flat, 1–1.5 cm long, 2-ranked on annual branchlets, pinnate, turning dark reddish brown in autumn. Pollen cones globose or ovoid, ca. 2.5 cm in diam., pendulous, pale yellowish brown at maturity, covered with white powder. Pollination Apr, seed maturity Oct.

# 墨西哥落羽杉 | *Taxodium mucronatum*

**别名：** 尖叶落羽杉　　　　　　　　　　**科属：** 杉科落羽杉属

**形态特征：** 常绿或半常绿大乔木，高达40 m。树干基部膨大，树冠阔圆锥形；树皮黑褐色，长条状脱落；小枝微下垂，侧生短枝螺旋状散生，在第2年春季脱落。叶条形，扁平，长1～2 cm，紧密排列成2列，翌年早春与小枝一起脱落。球果卵圆形，长1.5～2.5 cm。花期春季，秋后球果成熟。

**习性、应用及产地分布：** 喜温暖，耐寒性差，耐水湿，对碱性土适应力较强。原产于墨西哥及美国西南部海拔300～1 600 m地区，中国长江流域低湿河网地区造林绿化树种。

**Description:** Large trees, evergreen or semievergreen, up to 40 m tall. Crown broadly conical, trunk swollen at base; bark dark brown, peeling off in long strips; branchlets slightly pendulous, lateral ones spirally arranged, scattered, deciduous in early spring of following year. Leaves linear, flat, 1–2 cm long, 2-ranked, deciduous with branchlets in early spring of following year. Seed cones ovoid, 1.5–2.5 cm long. Pollination spring, seed maturity after autumn.

# 池杉 | *Taxodium distichum* var. *imbricatum*

**别名：** 池柏、沼杉　　　　　　　　　　　**科属：** 杉科落羽杉属

**形态特征：** 落叶乔木，高达25 m。树冠较窄，呈尖塔形；树皮褐色，纵裂成长条片脱落；树干基部膨大；1年生小枝绿色，细长，常略向下弯垂，2年生小枝褐红色。叶多钻形，略内曲，常在枝上螺旋状伸展，下部多贴生小枝，长4～10 mm。球果圆球形，长2～4 cm，熟时黄褐色。花期3—4月，球果10—11月成熟。

**习性、应用及产地分布：** 强阳性树种。极耐水湿，抗风力强，速生，是江南水网地区重要的防护林树种。树形优美，枝叶秀丽，是观赏价值很高的园林树种。适宜植作长江流域及珠江三角洲等平原地区、水库附近防护林。产于美国弗吉尼亚州南部等低湿地，中国南京、杭州、武汉等地多有引种。

**Description:** Trees, deciduous, up to 25 m tall. Crown relatively narrow and pyramidal; bark brown, peeling off in long strips, trunk swollen and buttressed at base; 1st-year branchlets green, soft and thin, slightly pendulous, 2nd-year branchlets reddish brown. Leaves mostly subulate, slightly incurved, usually spirally arranged on branchlets, 4–10 mm. Seed cones globose, 2–4 cm long, yellowish brown at maturity. Pollination Mar–Apr, seed maturity Oct–Nov.

# 水杉 | *Metasequoia glyptostroboides*

**科属：** 杉科水杉属

**形态特征：** 落叶巨乔木，高达40 m。树冠尖塔形，老则为广圆头形；树干基部常膨大；树皮灰褐色或深灰色，裂成条片状脱落；大枝近轮生，小枝近对生。叶条形，长1～1.7 cm，柔软，近无柄。雄球花卵圆形，单生或圆锥花序状生于叶腋或苞腋；雌球花单生于侧枝顶端。球果近球形，长1.8～2.5 cm，具长柄，下垂；种鳞极薄，透明；种鳞盾形木质，每个种鳞具种子5～9粒。花期2月，球果10—11月成熟。

**习性、应用及产地分布：** 耐水湿，少病虫害，抗有毒气体。树干通直挺拔，树形壮丽，叶色翠绿，秋叶棕褐色，季相变化明显，是重要的庭院观赏树，可于公园、庭院、草坪、绿地、建筑物前孤赏，效果均佳。仅1种，有"活化石"之称。中国特产，分布于重庆、湖南、湖北等山地，北京以南有栽培。国家一级重点保护植物。

**Description:** Known as the "living fossil". Large trees, deciduous, up to 40 m tall. Crown pyramidal, finally broadly conical; trunk buttressed at base; bark dark reddish brown or dark gray, fissured and exfoliating; branchlets opposite. Leaves linear, 1−1.7 cm, soft and subsessile. Pollen cones ovoid, female cones grows on top of branches of lateral branchlets. Seed cones subglobose, 1.8−2.5 cm, pendulous, seed scales extremely thin, transparent, each bearing 5−9 seeds. Pollination Feb, seed maturity Oct−Nov.

## 金叶水杉 | *Metasequoia glyptostroboides* 'Gold Rush'

**科属：** 杉科水杉属

**形态特征：** 落叶乔木，高达40 m。树皮红褐色，羽状复叶呈现金黄色。雄球花卵圆形，单生或圆锥花序状生于叶腋或苞腋；雌球花单生于侧枝顶端。球果近球形，长1.8～2.5 cm，具长柄。形态特征与水杉相似，是水杉的栽培品种。

**习性、应用及产地分布：** 树形端庄、稳重，生长快，性强健。中国北京、江苏、上海、浙江、四川、河南等地均有栽培引种。

**Description:** Trees, deciduous, up to 40 m tall. Bark reddish brown, leaves pinnate golden. Pollen cones ovoid, female cones singly fascicles, terminal on lateral branchlets. Seed cones subglobose, 1.8–2.5 cm, with a long handle. Morphological characteristics similar to *Metasequoia glyptostroboides*, is a cultivar of *Metasequoia glyptostroboides*.

## 台湾杉 | *Taiwania cryptomerioides*

**科属：** 杉科台湾杉属

**形态特征：** 常绿大乔木，高达75 m。树皮不规则条状剥落，内皮红褐色。叶螺旋状互生，鳞状锥形，横截面四棱形，基部两侧扁，长0.6～2 cm。球果长1.5～2.2 cm，种子长而扁，两侧有翅。

**习性、应用及产地分布：** 幼树耐荫，大树喜光；生长较快。产于中国台湾，云南、贵州和湖北也有分布。国家一级重点保护植物。

**Description:** Large trees, evergreen, up to 75 m tall. Crown conical or broadly rounded; bark cracking into long, irregular flakes, inner bark reddish brown. Leaves 0.6–2 cm long. Seed cones 1.5–2.2 cm long. Seeds narrowly elliptic, or narrowly elliptic-obovate with wings.

# 5. 柏科

## 金塔侧柏 | *Platycladus orientalis* 'Beverleyensis'

**科属**：柏科侧柏属

**形态特征**：常绿小乔木，树冠尖塔形。新叶金黄色，后渐变为黄绿色；枝叶扁平，排成一平面，两面同型；鳞叶二型，中央叶倒卵状菱形，两侧叶舟形。雌雄同株，阔卵形球果多单生于枝顶，长1.2～2.5 cm。花期3—4月，球果当年9—10月成熟。

**习性、应用及产地分布**：喜光，耐干旱瘠薄，不耐水涝；浅根系，侧根发达，寿命长。常用于庭院观赏，中国北京、南京、杭州等地园林绿地中常有栽培。

**Description:** Trees, evergreen, crown pyramidal. New leaves golden yellow, old leaves yellow-green; branches and leaves compressed; leaves 2 types, facial leaves rhomboid, lateral leaves boat-shaped. Pollen cones yellowish green, ovoid, 2–3 mm in diam. Seed cones usually born singly, terminal, 1.2–2.5 cm long. Pollination May-Apr, seed maturity Sep–Oct. Cultivated broadly in gardens of Beijing, Nanjing, Hangzhou, etc.

## 蓝冰柏 | *Cupressus glabra* 'Blue Ice'

**科属：** 柏科柏木属

**形态特征：** 为绿干柏的栽培品种。常绿乔木，树冠圆锥状。鳞叶交叉对生，蓝绿色，被白粉，背部有腺点，芳香。

**习性、应用及产地分布：** 喜温暖、湿润气候，耐瘠薄，抗性强；生长迅速，寿命长，根系发达，耐修剪。树姿优美，可孤植或丛植，是圣诞树的首选。原产于美国，中国河南、江苏、上海栽培较多。

**Description:** Trees, evergreen, crown conical. Scale leaves bluish green, glaucous, with a conspicuous abaxial gland, back with glandular point, fragrant.

# 圆柏 | *Sabina chinensis*

**别名：** 桧柏　　　　　　　　　　**科属：** 柏科圆柏属

**形态特征：** 大乔木，高20 m。树冠幼时呈尖塔形，老时呈广卵形。叶二型，鳞叶与刺叶兼有；鳞叶小，交互对生，刺叶3枚轮生，基部下延，无关节，有白色气孔带；雌雄异株，稀同株。球果近圆球形，被白粉；种子卵形，稍扁。花期4月下旬，球果翌年10—11月成熟。

**习性、应用及产地分布：** 温带阳性树种。喜温凉、温暖气候，对土壤要求不严。深根性，侧根发达；耐修剪，易整形；生长速度中等，寿命极长。是中国传统的园林树种，早在秦汉时期就已栽培应用，与古典建筑相得益彰。宜植于亭台、庭院或林缘、草坪边缘，以增加层次感。产于中国内蒙古及沈阳以南，南达两广（广东、广西的合称）北部；朝鲜、日本亦有分布。

**Description:** Large trees, evergreen, up to 20 m tall. Crown pyramidal when young, becoming broadly ovoid with age. Leaves 2 types, scalelike and spinelike, base decurrent, with white stomatal bands. Dioecious, seed cones subglobose, glaucous. Seeds ovoid, slightly flattened. Pollination Apr, seed maturity Oct–Nov of following year.

# 龙柏 | *Sabina chinensis* 'Kaizuka'

**科属：** 柏科圆柏属

**形态特征：** 为圆柏的栽培品种。常绿乔木。分枝低，侧枝短而环抱主干，大枝如龙柱扭曲上升，小枝密；多鳞叶，幼时鲜黄绿色，老时灰绿色。

**习性、应用及产地分布：** 喜光，抗烟尘及有害气体能力较强，耐修剪，易整形。常用于公园、庭园、绿篱整形修剪，公路隔离带，或作为庭院观赏树。中国长江流域各大城市均有栽培。

**Description:** Trees, evergreen. Lateral branchlets short, branchlets dense and twisted upward. Leaves yellow-green when young, becoming gray-green with age.

## 金星球桧 | *Sabina chinensis* 'Aureo-globosa'

**科属：** 柏科圆柏属

**形态特征：** 为圆柏的栽培品种。常绿灌木或小乔木。树冠近球形或塔形，枝密生。叶二型，鳞叶与刺叶兼有，被白粉；鳞叶小，交互对生，刺叶3枚轮生；枝端新萌发幼叶常金黄色，后渐变为绿色。

**习性、应用及产地分布：** 喜光，不耐水涝，耐修剪，适应性强。庭园观赏树，可行植、列植、丛植。中国各地均有栽植。

**Description:** Shrubs or trees, evergreen. Crown subglobose or pyramidal, dense. Leaves 2 types: scalelike and needlelike, glaucous, needlelike leaves in whorls of 3; newly emerged terminal leaves usually golden-yellow, becoming green with age.

# 日本花柏 | *Chamaecyparis pisifera*

**别名：** 花柏、五彩柏　　　　　　　　**科属：** 柏科扁柏属

**形态特征：** 常绿大乔木。树冠尖塔形；树皮深灰色或红褐色，大枝水平开展，生鳞叶，小枝扁平，近平行于地面。两侧鳞叶先端锐尖，上面暗绿色，下面被白粉显著，具不明显的腺点。球果暗褐色，径 6～8 mm。种子三角状卵形，具棱脊，两侧有宽翅。花期 7—8 月，球果 10—11 月成熟。

**习性、应用及产地分布：** 中性树种。喜温凉、湿润气候；根系发达，耐修剪。常密植作绿篱或绿墙。原产于日本，中国各大城市有引种。

**Description:** Large trees, evergreen. Crown pyramidal; bark dark gray or reddish brown, branches horizontally spreading. Lateral leaves acute at apex, facial leaves dark green, glaucous back, with obscure abaxial glands. Seed cones dark brown, 6–8 mm in diam. Seeds narrowly obovoid to transversely ellipsoid with broad wings. Pollination Jul–Aug, seed maturity Oct–Nov.

# 金线柏 | *Chamaecyparis pisifera* 'Filifera Aurea'

**科属：**柏科扁柏属

**形态特征：**常绿乔木。为日本花柏的栽培品种。外形如线柏，小枝及叶为金黄色，小枝细长而下垂，鳞叶紧贴。雌雄同株，雄球花椭圆形，雌球花单生于枝顶；球果当年成熟，球形，种鳞木质。种子卵圆形，微扁。花期3月，球果11月成熟。

**习性、应用及产地分布：**喜光，耐半阴，抗寒耐旱；幼苗期生长缓慢，耐修剪。金线柏枝叶细柔，姿态婆娑，园林中孤植、丛植、群植均宜。原产于日本，中国东部沿海各大城市均有栽培。

**Description:** Trees, evergreen. Branchlets and leaves golden yellow, branchlets long and pendulous. Monoecious. Seed cones globose, cone scale ligneous. Seeds ovoid, slightly compressed. Pollination Mar, seed maturity Nov.

# 6. 罗汉松科

## 罗汉松 | *Podocarpus macrophyllus*

**别名**：罗汉杉、土杉　　　　　　　　　　**科属**：罗汉松科罗汉松属

**形态特征**：常绿乔木。叶条状披针形，长7～12 cm，先端尖，螺旋状散生，两面中脉隆起，无侧脉。雄球花葇荑状，单生或簇生于叶腋；雌球花多1～2簇生于叶腋，亦有少数生于短枝枝顶，具柄。种子完全被肉质假种皮包被，着生在肉质或非肉质的种托上。种子成熟时假种皮紫黑色，被白粉，肉质种托紫红色，宛如身披袈裟打坐参禅的罗汉，故名。花期4—5月，种子8—9月成熟。

**习性、应用及产地分布**：亚热带阴性树种。喜温暖、湿润和半阴环境，耐海潮风；耐修剪，寿命长。抗病虫害；树形古雅苍劲，公园、庭院、寺庙多有种植。产于中国长江流域以南。

**Description:** Trees, evergreen. Bark gray or grayish brown, peeling off in thin flakes. Leaves linear-lanceolate, 7–12 cm long, apex acuminate, spirally arranged, midvein prominently raised adaxially, slightly raised abaxially. Pollen cones spikelike, solitary or in clusters; female cones in clusters of 1–2 on axil, minority terminal, pedunculate. Receptacle red or purplish red when ripe, columnar. Epimatium purplish black when ripe, covered with white powder. Pollination Apr–May, seed maturity Aug–Sep.

## 竹柏 | *Nageia nagi*

**别名：** 椤树、山杉　　　　　　**科属：** 罗汉松科竹柏属

**形态特征：** 常绿大乔木，高20～30 m。树冠圆锥形；树皮褐色，平滑，薄片状脱落。叶交叉对生，厚革质，有光泽，宽披针形或椭圆状披针形，长3.5～9 cm，宽1.5～2.5 cm，先端渐尖，基部楔形，无明显中脉，具多数并列细脉。雌雄异株。种子圆球形，径1.2～1.5 cm，成熟时肉质假种皮暗紫色，被白粉。花期3—5月，种子10—11月成熟。

**习性、应用及产地分布：** 亚热带及热带阴性树种，不耐修剪。"叶与竹类，致理如柏，以状得名"。树冠浓郁，终年苍翠，叶形奇异，为中国南方和日本特有的观叶树种，适宜作庭荫树、行道树，亦适宜盆栽观赏。产于中国岭南以南地区海拔800～900 m的山地林中。

**Description:** Large trees, evergreen, up to 20–30 m tall, dioecious. Crown conical; bark brown, smooth, peeling in small, thin flakes. Leaves opposite, decussate, leathery, glossy, ovate-lanceolate or elliptic-lanceolate, 3.5–9 cm × 1.5–2.5 cm, apex acuminate, base cuneate, midvein invisible, several parallel veins. Seeds globose, 1.2–1.5 cm in diam., aril dark purple when ripe, powdery. Pollination Mar–May, seed maturity Oct–Nov.

# 7. 三尖杉科

## 三尖杉 | *Cephalotaxus fortunei*

**别名：** 三尖松、山榧树　　　　　　　　　　**科属：** 三尖杉科三尖杉属

**形态特征：** 常绿乔木，高20 m。树冠广圆形；树皮褐色或红褐色，片状脱落；枝细长，稍下垂。叶在小枝上排为2列，披针状条形，微弯，长4～13 cm，宽3.5～4.5 mm，先端渐尖，基部楔形，叶背具2条白色气孔带，比边带宽3～5倍。雄球花径约1 cm；种子椭圆状卵形，长3.5～4.5 cm，假种皮紫红色。花期4月，种子翌年8—10月成熟。

**习性、应用及产地分布：** 亚热带阴性树种。不耐寒，对环境适应性强；生长较慢，耐修剪。植株含多种生物碱，可治疗恶性肿瘤。种子成熟时绿叶红果相映成趣，可植为园景树观赏。是国家生态建设工程首选树种，可营建水土保持林和水源涵养林，亦可作室内观赏盆景。中国特产，分布于秦岭、大别山至华南北部。

**Description:** Trees, evergreen, up to 20 m tall. Crown broadly ovoid; bark brown or reddish brown, peeling in strips; branchlets slightly pendulous. Leaves linear-lanceolate, falcate or straight, 4–13 cm × 3.5–4.5 mm, apex acuminate, base cuneate, 2 stomatal lines abaxially, 3–5 times wider than marginal bands. Pollen cones ca.1 cm in diam. Seeds ellipsoid-ovoid, 3.5–4.5 cm long, aril purple-red. Pollination Apr, seed maturity Aug–Oct of following year.

# 8. 红豆杉科

## 南方红豆杉 | *Taxus wallichiana* var. *mairei*

**别名：** 美丽红豆杉　　　　　　　　　　**科属：** 红豆杉科红豆杉属

**形态特征：** 常绿乔木。为红豆杉的变种，叶片较原种长而宽。叶镰形弯曲，叶长2.5～4 cm，宽3～4 mm，叶背有两条黄绿色气孔带。种子卵圆形坚果状，成熟时被鲜红色杯状肉质假种皮所包被。

**习性、应用及产地分布：** 阴性树种。幼苗期生长缓慢，寿命达数百年。植株含抗肿瘤的紫杉醇。树体高大，树形端正，针叶终年翠绿，与成熟时鲜红的假种皮交相辉映，十分美丽，是优良的园林观赏树种，可孤植、群植或列植于假山石旁或疏林下。中国特有种，分布于长江流域以南。

**Description:** Trees, evergreen. Leaves falcate, 2.5–4 cm × 3–4 mm, 2 yellow green stomatal lines abaxially. Seed cones ovoid, surrounded by bright red goblet fleshy aril at maturity.

# 东北红豆杉 | *Taxus cuspidata*

**别名**：紫杉　　　　　　　　　　**科属**：红豆杉科红豆杉属

**形态特征**：常绿乔木，高达20 m。树冠阔卵圆形；树皮赤褐色，呈片状剥裂；大枝近水平伸展。叶在枝条上螺旋状排列，叶条形，直或微弯，长1～2.5 cm，先端常突尖。种子坚果状，卵形或三角状卵形，微扁，长6 mm，赤褐色，假种皮深红色。花期5—6月，种子9—10月成熟。

**习性、应用及产地分布**：温带及寒带阴性树种。耐寒。浅根性，侧根发达；生长缓慢，耐修剪，寿命极长。树形端正，可孤植、对植或群植为绿篱或整形修剪为各种雕塑物式样，宜作高山园、岩石园材料或盆栽。产于中国东北长白山林中，江苏、江西、上海等地有栽培。

**Description:** Trees, evergreen, up to 20 m tall. Crown broadly ovoid; bark reddish brown, peeling in strips, branchlets horizontal. Leaves radially spreading on long branchlets, linear, straight or slightly curved, 1–2.5 cm long, apex usually shortly mucronate. Seeds ovoid or trigonous-ovoid, slightly compressed, 6 mm long, aril dark red. Pollination May–Jun, seed maturity Sep–Oct.

# 榧树 | *Torreya grandis*

**科属：** 红豆杉科榧树属

**形态特征：** 常绿乔木，高可达30 m。树皮淡黄灰色或灰褐色，不规则纵裂，1年生枝绿色。叶在小枝上排成2列，条形或条状披针形，长1.1～2.5 cm，坚硬，先端刺尖，中脉不明显，叶背中脉两侧具2条浅黄色气孔带。雄球花单生于叶腋，雌球花双生于叶腋，胚珠生于漏斗状珠托上。种子核果状，全部包于肉质假种皮中，翌年秋季成熟。

**习性、应用及产地分布：** 亚热带中性树种。生长缓慢，寿命长；少病虫害，抗烟尘能力强。树干高大挺拔，树冠繁密，枝叶葱绿，适于庭院、草坪及道路两侧栽植。产于中国华东地区。国家二级重点保护植物。

**Description:** Trees, evergreen, up to 30 m tall. Bark light yellowish gray or grayish brown, with irregular vertical fissures, 1st-year branchlets green. Leaves linear or linear-lanceolate, 1.1～2.5 cm long, hard, apex spinescent, midvein indistinct adaxially, with 2 pale yellow stomatal bands abaxially. Seed maturity autumn of following year.

# 下篇
# 被子植物

# 9. 杨梅科

## 杨梅 | *Myrica rubra*

**科属：** 杨梅科杨梅属

**形态特征：** 常绿乔木，高可达15 m，雌雄异株。树冠整齐，近球形；树皮黄灰色，老时浅纵裂。叶革质，叶柄短，多集生于枝顶，长椭圆状倒卵形至倒披针形，长6～16 cm，先端圆钝，基部狭楔形，表面有光泽，中脉明显，全缘或中上部有浅齿。雄花序穗状，腋生，紫红色，长1.5～3 cm；雌花序单生，长0.5～1.5 cm，红色至紫色。核果球状，径1.5～3 cm，深红色或紫色，外果皮肉质多汁。花期3—4月，果期6—7月。

**习性、应用及产地分布：** 喜温暖、湿润气候，具有一定的耐寒性。对二氧化硫、氯气等抗性较强。树冠圆整，枝繁叶茂，初夏红果累累，红绿交辉，是优良园林绿化树种。孤植、丛植于草坪、庭院及墙隅或列植于路旁均可。分布于中国长江流域及以南地区，其中浙江省栽培最多。

**Description:** Trees, evergreen, up to 15 m tall, dioecious. Crown subglobose; bark yellow-gray, aging fissured. Leaves leathery, petiole short, elliptic-obovate, 6–16 cm long, apex obtuse to acute, base cuneate, glossy, midvein distinct, margin entire or serrate in apical 1/2. Male spikes in leaf axils, purple-red, 1.5–3 cm; female spikes solitary in leaf axils, 0.5–1.5 cm long, red to purple. Drupe globose, 1.5–3 cm in diam., dark red or purple, edible. Fl. Mar–Apr, fr. Jun–Jul.

# 10. 胡桃科

## 野核桃 | *Juglans mandshurica*

**科属：** 胡桃科胡桃属

**形态特征：** 乔木，高可达25 m，幼枝密被腺毛及柔毛。奇数羽状复叶，小叶9～17枚，卵状长椭圆形至矩圆形，缘有细锯齿，叶片有星状毛，叶背密被短柔毛；托叶痕猴脸状。雄花序长20～30 cm，核果卵形。花期4—5月，果期8—10月。

**习性、应用及产地分布：** 喜光，深根性。在园林中可作道路绿化植物，起防护作用。产于中国中部、东部及西南地区。

**Description:** Trees, up to 25 m tall, branchlets densely covered with glandular hairs and pubescent. Leaves odd-pinnate, with 9–17 leaflets; leaflets long elliptic to ovate-elliptic, serrulate, abaxially slightly pubescent. Male spikes 20–30 cm. Nuts ovoid. Fl. Apr–May, fr. Aug–Oct.

## 枫杨 | *Pterocarya stenoptera*

**科属：** 胡桃科枫杨属

**形态特征：** 落叶乔木，高可达30 m。幼树皮红褐色，平滑，老树皮浅灰色至深灰色，纵裂；小枝具片状髓心。奇数或偶数羽状复叶，具小叶10～16枚，顶生小叶有时不发育而成假偶数羽状复叶，叶轴具叶质窄翅，小叶有锯齿。雄性葇荑花序长6～10 cm，单生于去年生枝叶腋；雌性葇荑花序长10～15 cm，单生于新枝顶端。果序下垂，长20～30 cm；坚果近球形，径6～7 mm，具2条长圆形斜展阔翅，翅长1.2～2 cm。花期4—5月，果期8—9月。

**习性、应用及产地分布：** 温带南部及亚热带强阳性树种。喜温暖、湿润气候，也较耐寒。耐湿性强，对土壤要求不严。深根性，对有毒气体具一定抗性。树冠宽广，枝叶茂密，果序随风飘动，优美动人，多作为庭荫树、行道树。广布于中国的华北、华中、华南和西南各省，为溪边及水湿地习见树种。

**Description:** Trees, deciduous, up to 30 m tall. Bark reddish brown when young, aging pale gray to dark gray, fissured. Leaves odd-pinnate with 10–16 leaflets; leaflets serrulate; rachis often winged. Male spikes 6–10 cm, solitary in last-year leaf axils; female spikes solitary on new shoots, 10–15 cm long. Fruiting spike pendulous, 20–30 cm long; nutlets subglobose, 6–7 mm in diam., wings linear, 1.2–2 cm long. Fl. Apr–May, fr. Aug–Sep.

# 11. 杨柳科

## 毛白杨 | *Populus tomentosa*

**别名**：大叶杨　　　　　　　　　　　**科属**：杨柳科杨属

**形态特征**：落叶乔木，高可达30～40 m。树干通直；树冠卵圆形或卵形，树皮青白色，皮孔菱形。叶三角状卵形，长10～15 cm，基部心形或平截，叶面暗绿色，叶背初时密生白绒毛，后渐脱落，边缘具波状缺刻或深锯齿；叶柄扁，顶端常具2～4枚腺体。雄花序长10～15 cm；雌花序长4～7 cm。果序长14 cm。蒴果小，三角形。花期3月，叶前开放；果期4—5月。

**习性、应用及产地分布**：温带阳性树种。喜凉爽、湿润气候。深根性，生长较快。耐烟尘，抗污染。树体高大挺拔，姿态雄伟，树干银白，叶大荫浓，适应性强，常用作行道树、庭荫树或用于营造防护林。中国特有种。分布广，北起辽宁南部、内蒙古，南至长江流域，以黄河中下游为分布中心。

**Description:** Trees, deciduous, up to 30−40 m tall. Crown conical to ovoid-globose, trunk erect, bark grayish green to grayish white. Leaves deltoid-ovate, 10−15 cm long, base cordate or truncate, adaxially dark green, abaxially densely tomentose, margin with deeply coarse or sinuate teeth, petiole laterally flattened, usually with 2−4 glands. Male catkins 10−15 cm. Female catkins 4−7 cm. Capsules deltoid. Fl. Mar, fr. Apr−May.

# 中华红叶杨 | *Populus deltoids* 'Zhonghua hongye'

**别名：**红叶杨、中红杨、变色杨　　　　**科属：**杨柳科杨属

**形态特征：**落叶大乔木。单叶互生，叶片三角状卵形，长12～25 cm，缘有锯齿，叶背灰绿色，叶柄与叶脉始终为红色。叶片季相变化明显，三季四变，早春展叶呈玫瑰红色，7～9月变为紫绿色，10月为暗绿色，11月后变为杏黄或金黄色。

**习性、应用及产地分布：**抗性强、适应性广、耐旱涝、耐低温。色泽亮丽诱人，为世界所罕见，观赏价值颇高。树干通直圆满、挺拔，生长迅速，是园林及道路绿化的优选树种。近年来在城市绿化中广泛种植。本种为芽变品种，在中国华南、西南及东北地区有栽培种植。

**Description:** Trees, deciduous. Leaf blades alternate, deltoid-ovate, 12–15 cm long, margin serrate, abaxially greyish-green, petiole and vein red. Leaf blade changes with seasons, prevernal rose-red, Jul to Sep violet-green, Oct dark green, after Nov apricot or auratus.

# 加杨 | *Populus × canadensis*

**别名：** 加拿大杨、加拿大白杨、美国大叶白杨　　**科属：** 杨柳科杨属

**形态特征：** 落叶乔木，干直，高可达30 m以上。树冠开展呈卵圆形，树皮深纵裂；小枝稍具棱脊，髓心近五角形。叶近三角形，长7～10 cm，先端渐尖，基部截形，两面无毛，边缘半透明，具圆钝齿；叶柄扁平而长，带红色。雄花序长7～15 cm，雄蕊15～25枚，苞片丝状深裂；雌花序具花40～50朵，柱头4裂，花盘全缘。果序长2～7 cm。蒴果卵圆形，2～3瓣裂。花期4月，果期5—6月。

**习性、应用及产地分布：** 温带阳性树种。喜暖热气候，耐寒。适生于湿润而排水良好的冲积土，对水涝、盐碱和瘠薄有一定抗性。树体高大，树冠宽阔，叶片大而具光泽，夏季绿荫浓密，秋叶金黄，是华北及江淮平原最常见的绿化树种之一。适宜作行道树、庭荫树及防护林树种。除广东、云南和西藏外，中国各省区均有栽培，其中以华北、东北及长江流域最多。

**Description:** Trees, deciduous, more than 30 m tall. Crown ovoid-globose, trunk erect; bark deeply furrowed. Leaf blades deltoid-ovate, 7–10 cm long, apex acuminate, base truncate, glabrous, margin crenate, translucent; petiole long, laterally flattened. Male catkin 7–15 cm long, bracts greenish lacerate; female catkin with 45–50 flowers. Female flower: stigma 4-lobed, margin entire. Infructescence 2–7 cm long. Capsule ovoid, 2- or 3-valved. Fl. Apr, fr. May–Jun.

# 垂柳 | *Salix babylonica*

**别名：** 水柳、垂丝柳、清明柳　　**科属：** 杨柳科柳属

**形态特征：** 高大落叶乔木，高可达18 m。树冠倒广卵形；树皮灰褐色，不规则开裂；小枝淡黄褐色，细长，下垂，髓心圆形。叶条状披针形，长8～16 cm，宽0.5～1.5 cm，无毛或幼叶微被毛，边缘具细锯齿，叶柄长0.5～1.5 cm。葇荑花序直立，无花被；雄花序长1.5～3 cm，雄蕊2枚，花丝分离，苞片具腺体2枚；雌花序长2～3 cm，苞片具腺体1枚。花期3—4月，果期（飞柳絮）4—5月。

**习性、应用及产地分布：** 温带及亚热带阳性树种。极耐水湿，浅根性，根系发达；萌芽力强。生长迅速；寿命较短，30年后渐趋衰老。常植于河岸及湖池边随风飘舞，依依拂水，别有风致，亦可用作行道树、庭荫树、固岸护堤树及水土保持树种。产于中国长江流域及其以南平原地区。

**Description:** Trees to 18 m tall. Bark gray-brown, irregularly furrowed; branchlets pale yellow, slender, pendulous. Leaf blade linear-lanceolate, 8–16 cm × 0.5–1.5 cm, both surfaces glabrous or slightly pilose, margin serrate, petiole 0.5–1.5 cm long. Male catkin 1.5–3 cm long, stamens 2, bracts with 2 glands; female catkin 2–3 cm, bracts with 1 gland. Fl. Mar–Apr, fr. Apr–May.

# 龙爪柳 | *Salix matsudana* 'Tortuosa'

**科属：** 杨柳科柳属

**形态特征：** 高可达12 m，枝条自然扭曲向上，其他同原种。常用于栽培观赏，生长势较弱，易衰老。

**Description:** Branchlets twisted upward. Other specialities are similar to the original species.

# 银芽柳 | *Salix × leucopithecia*

**别名：** 银柳、棉花柳　　　　　　　　　**科属：** 杨柳科柳属

**形态特征：** 灌木，高 2～3 m。分枝稀疏，小枝绿褐色，具红晕，幼时被绢毛。叶半革质，长椭圆状披针形，长 9～15 cm，下面密被白色柔毛，边缘有细锯齿。花芽肥大，芽具 1 片紫红色且富光泽的苞片，冬季及早春先于叶开放；雄花序椭圆状圆柱形，长 3～6 cm，初开时芽鳞疏展，盛开时苞片脱落，即露出密被的银白色绢毛（花序），形似毛笔，颇为美观。花期 12 月至翌年 2 月。

**习性、应用及产地分布：** 喜光，颇耐寒。耐涝，适生于水边及肥沃湿润的土壤。观芽植物。花序萌发时十分美观，多在春节前后与一品红、水仙配置瓶插观赏，极富东方艺术之神韵。产于日本，中国上海、南京、杭州、北京等地有栽培。

**Description:** Shrubs, 2–3 m tall. Branchlets green-brown. Leaf blade 9–15 cm long, abaxially densely white tomentose, margin serrulate. Male catkins elliptic-cylindrical, 3–6 cm long. Fl. Dec to Feb of following year.

## 彩叶杞柳 | *Salix integra* 'Hakuro Nishiki'

**科属：** 杨柳科柳属

**形态特征：** 灌木。小枝无毛。叶近对生或对生，有时3叶轮生，椭圆状长圆形，长2～5 cm，先端短渐尖，基部圆或微凹，全缘或上部有尖齿，嫩叶白色带粉红，后渐变为绿色有黄白色斑纹，两面无毛。花先于叶开放，花序对生，稀互生，长1～2.5 cm，基部有小叶；雄蕊2枚，花丝基部合生，花药红紫色。蒴果长2～3 mm，有毛。花期5月，果期6月。

**习性、应用及产地分布：** 喜光，耐水湿，常生于河边和湿地，是固岸护坡的好树种。本种是杞柳的栽培变种，从加拿大引入，中国北京、华北、华东等地有栽培。

**Description:** Shrubs. Branchlets glabrous. Leaves subopposite or opposite, sometimes in whorls of 3 on shoots, leaf blade elliptic-oblong, 2–5 cm long, apex shortly acuminate, base rounded or retuse, margin entire or sharply dentate distally, white with pink when young, aging green, glabrous. Flowering precocious; male flower: stamens 2, connate, anthers red or dull red. Capsule 2–3 mm, pilose. Fl. May, fr. Jun.

# 红叶腺柳 | *Salix chaenomeloides* 'Variegata'

**别名**：红心柳、红叶柳　　　　　　　　**科属**：杨柳科柳属

**形态特征**：落叶小乔木。小枝红褐色，有光泽。叶椭圆形、卵圆形或椭圆状披针形，长4～8 cm，先端渐尖，基部楔形，两面无毛，叶背苍白或灰白色，具腺齿；叶柄长0.5～1.2 cm，先端有腺点，托叶半圆形或长圆形，早落，顶端新叶常为亮红色。花期4月，果期5月。

**习性、应用及产地分布**：耐寒、耐湿、抗病虫害，适应性强。可作为河边、湖边及景观区点缀。分布于中国辽宁南部，黄河中下游及长江中下游，河南、山东、陕西、安徽、江苏、浙江等地。

**Description:** Trees, deciduous. Branchlets reddish brown, shiny. Leaf blade elliptic, ovoid or ovoid-lanceolate, 4–8 cm long, apex acuminate, base cuneate, glabrous, abaxially white or gray white, petiole 0.5–1.2 cm. Fl. Apr, fr. May.

# 12. 桦木科

## 桤木 | *Alnus cremastogyne*

**别名**：水冬瓜树　　　　　　　　**科属**：桦木科桤木属

**形态特征**：落叶乔木，高可达30～40 m。树皮灰色，幼时光滑，老则斑状开裂；小枝较细，灰褐色。叶倒卵状椭圆形，长6～15 cm，叶面疏生腺点，叶背密生腺点，先端骤尖或锐尖，基部楔形或近圆形，缘疏生细齿；侧脉8～10对。雌雄花序均单生，雄花序长3～4 cm。果序单生于叶腋，矩圆形，长1～3.5 cm；果梗纤细，长4～8 cm，下垂。花期3月，果期8—10月。

**习性、应用及产地分布**：温带阳性树种。喜温暖气候，耐水湿。对土壤适应性较强。根系发达，生长迅速，常能飞籽成林。具根瘤，可改良土壤。枝叶繁茂，耐水湿性强，是优良护岸、固堤树种，常在水滨、湖边成行种植，颇具野趣。产于中国四川、贵州和陕西等地。

Description: Trees, deciduous, up to 30–40 m tall. Bark gray, smooth when young, aging furrowed; branchlets slender, gray-brown. Leaf blade obovate-oblong, 6–15 cm long, adaxially sparsely glandular, abaxially densely glandular, apex abruptly acute, base cuneate or subrounded, margin obscurely and remotely obtusely serrate; lateral veins 8–10 on each side of midvein. Male and female inflorescences both solitary, male inflorescence 3–4 cm. Fruit inflorescence axillary, oblong, 1–3.5 cm long, slender, 4–8 cm, pendulous. Fl. Mar, fr. Aug–Oct.

# 千金榆 | *Carpinus cordata*

**别名:** 千金鹅耳枥、穗子榆、金丝榆　　**科属:** 桦木科鹅耳枥属

**形态特征:** 中乔木,高18 m。树皮亮灰色,光滑;小枝棕色或橘黄色,具沟槽。叶厚纸质,卵形或椭圆形,长8~15 cm,宽4~5 cm,先端渐尖,基部斜心形,叶面深绿色,秋季变为金黄色,叶背沿脉疏被短柔毛,边缘具尖锐重锯齿;侧脉15~20对;叶脉凹。葇荑花序与叶同放;果序长5~12 cm,膜质果苞宽卵圆形,两侧近对称,中脉位于近中央。花期5月,果期9—10月。

**习性、应用及产地分布:** 温带树种。叶色翠绿,可用于公园、绿地、小区绿化,适合孤植于草地、路边或三五株点缀栽培观赏。产于中国东北、河南、陕西、湖北等省,朝鲜、日本、俄罗斯亦有分布,在欧洲部分地区构成森林。

**Description:** Trees to 18 m tall. Bark shiny gray, smooth; branchlets brown or yellow-brown, grooved. Leaves thick, papery, ovate or elliptic, 8–15 cm × 4–5 cm, apex acuminate, base unequally cordate, adaxially dark green, turning golden-yellow in autumn, abaxially sparsely villous along midvein, margin irregularly and doubly setiform serrate; lateral veins 15–20 on each side of midvein; fruit inflorescence 5–12 cm long, densely pubescent; bracts broadly ovate-oblong, bilaterally subsymmetric, midvein in the middle. Fl. May, fr. Sep–Oct.

## 13. 壳斗科

### 板栗 | *Castanea mollissima*

**别名：** 栗子、毛栗　　　　　　　　　**科属：** 壳斗科栗属

**形态特征：** 落叶乔木，高15 m。树冠扁球形；树皮深灰色，不规则深纵裂；小枝被灰色绒毛。叶长椭圆形，长11～17 cm，宽7 cm，先端渐尖，叶背具灰白色短柔毛，边缘具芒齿；侧脉10～18对。雄花序直立，长9～20 cm，花3～5朵聚生成簇；雌花序着生于雄花序的基部。壳斗球形，径6～8 cm，果苞针刺状，熟时开裂，内含1～3枚褐色坚果。花期5—6月，果期9—10月。

**习性、应用及产地分布：** 温带阳性树种。较耐寒，耐旱，忌积水；对土壤要求不严。深根性，耐修剪。树冠宽广，叶片较大，可结合干果生产孤植或群植于庭院或草地，亦可用于营造山区水土保持林。中国特产，广泛栽培于辽宁、河北、黄河流域及其以南各省区。

**Description:** Trees, deciduous, 15 m tall. Crown ellipsoid; bark dark gray, irregularly furrowed; branchlets with gray pubescence. Leaf blade elliptic-oblong, 11–17 cm × 7 cm, apex acuminate, abaxially gray-white villous, margin serrate; lateral veins 10–18 on each side of midvein. Male inflorescences 3–5, vertical, 9–20 cm; female inflorescences born at the base of male inflorescences. Cupule globose, 6–8 cm in diam., bracts spinelike, dehiscent at maturity, nuts usually 1–3 per cupule. Fl. May–Jun, fr. Sep–Oct.

# 钩栗 | *Castanopsis tibetana*

**别名**：钩栲、大叶锥栗　　　　　　　　**科属**：壳斗科锥栗属

**形态特征**：常绿乔木，高可达30 m。干皮大片状剥落。叶大而坚硬，长椭圆形，长15～30 cm，叶缘中部以上有锯齿，叶背密被锈色绒毛。种苞密生细长刺，坚果单生于种苞内。

**习性、应用及产地分布**：中性树种，喜湿润、肥沃土壤，萌芽力强。宜孤植或丛植于草坪；有较好隔音、防尘、抗有毒气体能力，可用作防护林。广布于中国长江以南大部分省区。

**Description:** Trees, evergreen, up to 30 m tall. Bark peeling in strips. Leaf blade large and hard, oblong, 15−30 cm long, margin serrate except basally entire, abaxially covered with rusty to yellowish brown, waxy scalelike trichomes. Cupule globose, bracts spinelike, entirely covering cupule, nuts solitary per cupule.

# 苦槠 | *Castanopsis sclerophylla*

**科属：** 壳斗科锥栗属

**形态特征：** 中乔木，高可达15 m。树皮深灰色，纵裂；幼枝绿色，常具棱沟。叶厚革质，长椭圆形或卵状椭圆形，长7～15 cm，宽3～6 cm，先端长渐尖，基部阔楔形或圆形，有时略不对称，下面被灰白色或浅褐色蜡质层，边缘中部以上具粗锐锯齿；侧脉10～14对；叶柄长1.5～2.5 cm。果序长8～15 cm；坚果生于球状壳斗内，外被环状排列的瘤状苞片，果实成串生于枝上。花期4—5月，果期9—11月。

**习性、应用及产地分布：** 喜光，幼年耐荫；喜温暖、湿润气候。栽培以深厚肥沃、排水良好的酸性和中性砂质壤土为佳。树冠浑圆，枝叶繁密，颇为美观，宜于大型公园或绿地、庭院群植，或作为花木的背景树。产于中国秦岭以南，以华东和华中为主产地。

**Description:** Trees up to 15 m tall. Bark dark gray, irregularly furrowed; young shoots green, slightly angulate. Leaf blade leathery, oblong or ovate-elliptic, 7–15 cm × 3–6 cm, apex acuminate, base rounded to broadly cuneate, and sometimes slightly asymmetrical, abaxially covered with a gray-white or pale brown waxy coating, margin from middle to apex coarsely serrulate; secondary veins 10–14 on each side of midvein; petiole 1.5–2.5 cm long. Infructescences 8–15 cm long; nuts born in globular cupules, outer by annular bracts, fruit clustered on branches. Cupule globose, enclosing the nut, bracts scalelike, 3- or 4-angled, arranged in annular umbones. Fl. Apr–May, fr. Sep–Nov.

# 青冈栎 | *Cyclobalanopsis glauca*

**别名**：青冈、青栲　　　　　　　　　　**科属**：壳斗科青冈属

**形态特征**：大乔木，高可达20 m；小枝无毛。叶卵状椭圆形或长椭圆形，长6～13 cm，宽2～5.5 cm，先端渐尖或尾尖，基部近圆形或宽楔形，叶背密被整齐平贴白色毛，老时渐脱落，叶中部以上有疏锯齿，叶柄长1～1.5 cm。雄花序长5～6 cm，雌花序长约1 cm。壳斗碗状，外具5～8条排列紧密的同心环带；坚果卵形或近球形，1/3被壳斗包围，顶端被柔毛。花期3—5月，果期10月。

**习性、应用及产地分布**：亚热带中性树种。喜光，耐干燥。深根性，直根系，萌芽力强。防风、防火效果显著。树形优美，四季常绿，生性强健，为优良的园林绿化乡土树种，宜丛植、群植于庭院、大型公园、荒坡或作为工矿地绿篱、防风林、防火林等。广布于中国长江流域以南各省区，朝鲜、日本、印度亦有分布。

**Description:** Trees to 20 m tall; branchlets glabrous. Leaf blade obovate to oblong-elliptic, 6–13 cm × 2–5.5 cm long, apex acuminate to somewhat caudate, base rounded to broadly cuneate, margin apical 1/2 remotely serrate, petiole 1–1.5 cm. Abaxially gray-white villous but glabrescent. Male inflorescences 5–6 cm, female inflorescences ca. 1 cm. Cupule bowl-shaped, bracts in 5–8 rings, crowded; nut ovoid or ellipsoid, enclosed 1/3 by cupule, glabrous or rarely hairy. Fl. Mar-May, fr. Oct.

# 麻栎 | *Quercus acutissima*

**别名**：栎树、橡树　　　　　　　　　　**科属**：壳斗科栎属

**形态特征**：落叶乔木，高可达30 m。树皮深褐色，深纵裂；幼枝被黄色柔毛，后渐脱落。叶长椭圆状披针形，长8～19 cm，宽2～6 cm，先端渐尖，基部圆形或宽楔形，幼时被短柔毛，老时近无毛，边缘具芒状锯齿；羽状侧脉直达齿端；叶柄长1～3 cm。雄花序为下垂的荑葇花序，簇生于叶腋；壳斗杯形，包被坚果1/2，小苞片钻形、扁条形，反曲；坚果椭圆形，径1.5～2 cm。花期3—4月，果期翌年9—10月。

**习性、应用及产地分布**：温带阳性树种。喜温暖、干燥气候，较耐寒。耐旱，不耐水湿。根系深，萌蘖力强，抗风、抗烟尘能力强。树形挺拔，枝叶浓密，叶色随季节变化明显，抗逆性强，适宜植为庭荫树或工矿厂区行道树及防护林。产于中国辽宁、华北、华东、华中、华南及西南，欧洲及亚洲的日本、朝鲜、越南、印度亦有分布。

**Description:** Trees, deciduous, up to 30 m tall. Bark dark brown, deeply furrowed; young branchlets yellowish gray tomentose, glabrescent with age. Leaf blade narrowly elliptic-lanceolate, 8–19 cm × 2–6 cm, apex acuminate, base rounded to broadly cuneate, tomentose when young, nearly glabrescent with age, margin with spiniform teeth; secondary veins extending to the teeth; petiole 1–3 cm. Cupules cupular, enclosing 1/2 of nut, bracts subulate to ligulate, reflexed; nut ellipsoid, 1.5–2 in diam. Fl. Mar–Apr, fr. Sep–Oct of following year.

# 栓皮栎 | *Quercus variabilis*

**别名：** 软木栎、粗皮栎　　　　　　　　**科属：** 壳斗科栎属

**形态特征：** 落叶乔木，高可达30 m。树皮木栓层很发达；小枝灰棕色，无毛。叶卵状披针形，长8～20 cm，宽2～8 cm，下面密被灰白色星状绒毛，边缘具芒状锯齿；羽状侧脉直达齿端；叶柄长1～3 cm。花序轴密被柔毛；壳斗常单生，杯状，包被坚果约2/3，小苞片钻形，反曲，有短毛。坚果径1.5 cm。花期3—4月，果期翌年9—10月。

**习性、应用及产地分布：** 温带、亚热带及热带阳性树种。耐低温，对土壤要求不严；主根发达，萌芽力强，寿命长。树形挺拔，枝叶浓密，可孤植、群植于庭院或街道。产于中国华北、西北、华东、华中、华南及西南地区，朝鲜、日本亦有分布。

**Description:** Trees, deciduous, up to 30 m tall. Bark with a well-developed cork layer; branchlets grayish brown, glabrous. Leaf blade ovate-lanceolate, 8−20 cm × 2−8 cm, abaxially densely grayish stellate tomentose, margin with spiniform teeth; secondary veins extending to the teeth; petiole 1−3 cm. Inflorescences densely pilose; cupule solitary, cupular, enclosing 2/3 of nut, bracts subulate, inflexed, pilose. Nut ca. 1.5 cm in diam. Fl. Mar−Apr, fr. Sep−Oct of following year.

## 沼生栎 | *Quercus palustris*

**科属：** 壳斗科栎属

**形态特征：** 落叶大乔木，高可达25 m。小枝褐色，无毛。叶倒卵形或椭圆形，长10～20 cm，宽7～10 cm，边缘具5～7条羽状深裂，裂片具细裂齿；叶柄长2.5～5 cm，叶色秋季变红。下垂的雄花序簇生，与叶同放。壳斗杯状形，径1.5～1.8 cm，包被坚果1/4～1/3，小苞片三角形，覆瓦状紧密排列，无毛而有光泽。坚果长椭圆形，径约1.5 cm。花期4—5月，果翌年秋季成熟。

**习性、应用及产地分布：** 温带阳性树种。喜温暖、湿润气候，极耐水湿，抗风能力强，耐空气污染。冠形优美，秋叶色彩斑斓，宜植于园林绿地观赏，为向阳河湖、湿地良好的绿化树种。产于美洲东部，中国山东、北京、辽宁等地有栽培。

**Description:** Trees, deciduous, up to 25 m tall. Branchlets brown, glabrous. Leaf blade obovate to elliptic, 10–20 cm × 7–10 cm, margin with 5–7 lobes on each side, lobes with fine serrated teeth; petiole 2.5–5 cm, turning red in autumn. Pendulous male inflorescences clustered with leaves. Cupule cupular, 1.5–1.8 cm in diam., enclosing 1/4–1/3 of nut, bracts triangular, crowded, glabrous, glossy. Nut narrowly ellipsoid, ca. 1.5 cm in diam. Fl. Apr–May, fr. autumn of following year.

## 弗吉尼亚栎 | *Quercus virginiana*

**别名：**强生栎　　　　　　　　　　**科属：**壳斗科栎属

**形态特征：**常绿乔木，高 10～15 m。树皮黑褐色，大枝平展，枝条柔韧。单叶互生，椭圆倒卵形，叶形多变，全缘或边缘具不规则刺状，略反卷；叶长 4～10 cm，表面有光泽。坚果包被于球状壳斗内，壳斗外被瘤状苞片。

**习性、应用及产地分布：**树冠拱形，树形优美。根系发达，萌蘖力强，是优良的城市观赏树种，用于城市绿化，在街道、公园、校园和球场用作遮荫树。原产于美国东南部的弗吉尼亚沿海平原。

**Description:** Trees, evergreen, up to 10–15 m tall. Bark dark brown, branches horizontal, flexible branches. Leaves alternate, oval to obovate, variable in shape, entire or margin irregularly spiny, slightly recurved, 4–10 cm long, shiny surface. Nuts encased in globular shells, covered with tuberculate bracts.

## 14. 榆科

### 榆树 | *Ulmus pumila*

**别名：** 白榆、家榆　　　　　　**科属：** 榆科榆属

**形态特征：** 大乔木，高可达25 m。树干直立，树冠圆球形；树皮暗灰色，不规则纵裂；枝多开展。叶卵状长圆形，长2～6 cm，宽1.2～3 cm，先端渐尖，基部偏斜，叶缘有不规则重锯齿或单锯齿。花簇生于枝上，翅果近圆形，长1～1.5 cm，果核位于翅果中部，熟时黄白色，俗称"榆钱"；果梗无毛。花期3—4月，果期4—6月。

**习性、应用及产地分布：** 阳性树种。适应性极强，耐寒、耐干旱；适生于肥沃湿润而排水良好的土壤，耐盐碱性强，深根性，侧根发达，萌芽力强，耐修剪。对烟尘及有毒气体抗性较强。枝叶稠密，是城乡重要的庭荫树、行道树。中国各省均有分布。

**Description:** Trees, deciduous, up to 25 m tall. Trunk erect, crown globose; bark dark gray, irregularly longitudinally fissured; branching development. Leaf blade elliptic-ovate, 2–6 cm × 1.2–3 cm, apex acuminate, base obliquely or symmetrically obtuse to rounded, margin simply or sometimes doubly serrate. Inflorescences fascicled cymes on second-year branchlets, samaras orbicular, 1–1.5 cm long, seed at center of samara, yellow-white at mature; stem glabrous. Fl. Mar–Apr, fr. Apr–Jun.

# 榔榆 | *Ulmus parvifolia*

**科属：** 榆科榆属

**形态特征：** 落叶乔木，高可达25 m。树冠广圆形；树皮灰褐色，呈不规则鳞片状剥落，内皮红褐色；1年生枝深褐色，密被柔毛，后渐脱落。单叶互生，卵状长椭圆形，长2～5 cm，宽1～2 cm，先端渐尖，基部偏斜，上面沿中脉被疏毛，下面脉腋有簇生毛，边缘具单锯齿。花3～6朵簇生于叶腋或排成簇状聚伞花序。翅果椭圆形或卵状椭圆形，长10～13 mm，凹缺被毛。花期8—9月；果期10—11月。

**习性、应用及产地分布：** 喜光，稍耐荫；喜温暖湿润气候，亦能耐-20℃短期低温。较耐干旱瘠薄，在酸性、中性和碱性土上均能生长。生长速度中等，寿命较长；深根性，抗风力强；萌芽力强，耐修剪。对二氧化硫等有毒气体及烟尘抗性较强。树姿优美，树皮斑驳，枝叶细密，在庭院中孤植、丛植为庭荫树或与亭榭、山石配置均很合适。产于中国华北、华东及西南各省平原、丘陵、山谷及坡地，日本、朝鲜亦有分布。

**Description:** Trees, deciduous, up to 25 m tall. Crown broadly orbicular; bark grayish brown, exfoliating into irregular scale-like flakes, inner bark reddish brown; 1st-year branchlets dark brown, densely pubescent. Leaves alternate, narrowly elliptic, 2–5 cm long, apex acuminate, base oblique, margin simply serrate. Inflorescences fascicled cymes, 3–6 flowered. Samaras elliptic to ovate-elliptic, 10–13 mm long. Fl. Aug–Sep, fr. Oct–Nov.

# 榉树 | *Zelkova serrata*

**科属：** 榆科榉属

**形态特征：** 落叶乔木，高可达35 m。树皮灰褐色至深灰色，树干通直不裂，老时呈不规则片状剥落。叶厚纸质，卵形至椭圆形，长3～10 cm，宽1.5～4 cm，先端渐尖或锐尖，基部近圆形、稍偏斜，叶面被糙毛，叶背密生灰色柔毛，边缘具整齐桃形锯齿；侧脉8～15对；叶柄粗短，被柔毛。雄花1～3朵，雌花及两性花单生。核果，上部微偏斜，径约4 mm。花期3—4月，果期9—11月。

**习性、应用及产地分布：** 喜光，适生于温暖、湿润气候及肥沃、排水良好的土壤，忌积水，不耐干旱贫瘠；深根性，抗风力强；耐烟尘，抗有毒气体。树形雄伟，绿荫浓密，枝细叶美，入秋叶色红艳，是江南地区重要的秋色叶树种。孤植、丛植或与其他常绿树种混植用作庭荫树，列植于林荫大道两侧及街道、公路两旁。产于中国秦岭、淮河以南至两广，西至西南。

**Description:** Trees, to 30 m tall, deciduous. Bark grayish white to grayish brown, exfoliating. Leaf blade subleathery, ovate to elliptic, 3–10 cm long, apex acuminate to caudate, base slightly oblique, rounded, adaxially hispid, abaxially densely gray pubescent, margin serrate to crenate; secondary veins 8–15 on each side of midvein. Petiole stout, pubescent. Male flowers 1–3 in clusters, female flowers solitary. Drupe subsessile, upper deflection, ca. 4 mm in diam. Fl. Mar–Apr, fr. Sep–Nov.

# 珊瑚朴 | *Celtis julianae*

**别名：** 大果朴　　　　　　　　　　　　　　**科属：** 榆科朴属

**形态特征：** 落叶乔木，高可达 30 m。树冠圆球形；树皮淡灰色，平滑。小枝、叶背及叶柄密被黄褐色绒毛。叶宽卵形至卵状椭圆形，长 6～12 cm，宽 3.5～8 cm，先端短渐尖或尾尖，基部近圆形，偏斜，中部以上或近全缘具浅锯齿，叶脉下凹。花单生于枝条上部叶腋。核果单生于叶腋，椭圆形至近球形，金黄至橙黄色；果柄粗壮，较叶柄长。花期 3—4 月，果期 9—10 月。

**习性、应用及产地分布：** 温带及亚热带树种。喜光，不耐寒，深根性；少病虫害，抗烟尘及有毒气体。树姿雄伟，冠大荫浓，入秋红果状如珊瑚，可孤植、丛植或列植为庭荫树、行道树及观赏树。产于中国长江流域，浙江、安徽南部、陕西南部、湖南西北部及贵州等地有栽培。

**Description:** Trees, deciduous, up to 30 m tall. Crown orbicular; bark light gray, smooth. Branchlets, abaxially and petiole densely brownish yellow pubescent. Leaf blade broadly ovate to ovate-elliptic, 6–12 cm × 3.5–8 cm, apex shortly acuminate to caudate-acuminate, base rounded and slightly oblique, margin finely toothed above middle to rarely subentire. Flowers solitary axillary. Drupe solitary, axillary, ellipsoid to globose, orange-yellow, pedicel stout, longer than petiole. Fl. Mar–Apr, fr. Sep–Oct.

# 小叶朴 | *Celtis bungeana*

**别名：** 黑弹树　　　　　　　　　　**科属：** 榆科朴属

**形态特征：** 落叶乔木，高可达10 m。树皮暗灰色，平滑，小枝通常无毛。叶卵形至卵状长椭圆形，长4～8 cm，先端长渐尖，基部宽楔形至近圆形，稍偏斜，中上部具不规则浅齿。果单生于叶腋，近球形，径6～8 mm，成熟时蓝黑色；果柄为叶柄长2倍或更长。花期4—5月，果期10—11月。

**习性、应用及产地分布：** 喜光，也较耐荫，耐寒，耐旱；深根性，萌蘖力强。枝叶繁茂，树形美观，宜作庭荫树及城乡绿化树种。产于中国东北、华北、长江流域及西南各省。

**Description:** Trees, deciduous, 10 m tall. Bark dark gray, smooth, branchlets glabrous. Leaf blade ovate to ovate-oblong, 4–8 cm, apex acute to acuminate, base lightly oblique, margin irregularly and narrowly ovate oblong or ovate. Drupe solitary, globose, 6–8 mm in diam., blackish blue when mature. Fl. Apr–May, fr. Oct–Nov.

## 青檀 | *Pteroceltis tatarinowii*

**科属**：榆科青檀属

**形态特征**：落叶乔木，高20 m。树皮灰色或深灰色，呈不规则长薄片状剥落，内皮淡灰绿色；小枝黄绿色，疏被短柔毛，皮孔明显。叶卵形，长3～10 cm，宽2～5 cm，先端具尾状渐尖，基部近圆形，稍偏斜，下面脉腋有簇毛，边缘具不整齐锯齿；基部三出脉，侧脉不直达齿端。花单性同株，生于叶腋；雄花簇生，雌花单生。坚果近圆形，先端凹缺，周围具薄翅，果柄细长。花期4月，果期7—8月。

**习性、应用及产地分布**：喜光，稍耐荫，较耐寒，耐干旱瘠薄；多生于石灰岩及钙质土壤；根系发达，萌芽性强，寿命长；病虫害少，对有毒气体有一定抗性。树体高大，树冠开阔，树皮灰白洁净，枝叶秀丽，秋季叶色金黄，果实美观，宜作庭荫树、行道树或与大型山石配置，以赏其秋季美景。产于中国辽宁大连以南广大地区。

**Description:** Trees, deciduous, up to 20 m tall. Bark gray or dark gray, peeling off in irregular long strips, inner bark pale gray-green; branchlets yellow-green, pubescent, lenticels distinct. Leaf blade ovate, 3–10 cm × 2–5 cm, apex caudate-acuminate, base oblique, rounded, margin irregularly serrate. Flowers axillary; male flowers fascicled, female solitary. Nut globose to oblong, apex notched, fruiting pedicel slender. Fl. Apr, fr. Jul–Aug.

# 15. 交让木科

## 交让木 | *Daphniphyllum macropodum*

**科属：** 交让木科交让木属

**形态特征：** 常绿乔木，高3～10 m。小枝粗壮，暗褐色，具明显圆形托叶痕，枝叶无毛。单叶互生，革质，长椭圆形至倒披针形，长14～25 cm，宽3～6.5 cm，先端渐尖，基部楔形至阔楔形，叶面具光泽，侧脉纤细而密，12～18对；新叶集生于枝端，叶柄紫红色，粗壮，长3～6 cm。雄花序长5～7 cm，雌花序长4.5～8 cm。果椭圆形，长约10 mm，先端具宿存柱头。花期3—5月，果期8—10月。

**习性、应用及产地分布：** 中性偏阴，喜温暖、湿润气候。可作为庭园观赏树。产于中国长江流域以南地区，日本、朝鲜也有分布。

**Description:** Trees up to 3–10 m tall, evergreen. Branchlets stout, bark dark brown, with orbicular leaf scars, leaves and branches are glabrous. Leaf blade leathery, oblong or oblanceolate, 14–25 cm × 3–6.5 cm, apex acuminate, base cuneate or broadly cuneate, shining adaxially; lateral veins slender and dense, 12–18 pairs, visible on both surfaces. Petiole, purplish red, stout, 3–6 cm long. Male inflorescence 5–7 cm, female inflorescence 4.5–8 cm. Drupe ellipsoidal, ca. 10 mm, style branches persistent. Fl. Mar–May, fr. Aug–Oct.

# 16. 杜仲科

## 杜仲 | *Eucommia ulmoides*

**科属：** 杜仲科杜仲属

**形态特征：** 落叶乔木，高20 m；树冠圆球形；树皮灰色，纵裂；枝、叶、果及树皮等处断裂后均有弹性胶丝相连。小枝光滑，具片状髓。叶卵状椭圆形，长6～16 cm，宽4～9 cm，先端渐尖，基部圆形或宽楔形，边缘有锯齿；叶脉下陷，叶背网脉明显，脉上有毛。花单性异株，生于当年枝基部。果长椭圆形，长3～3.5 cm，宽1～1.3 cm，扁平，两侧具窄翅。花期4—5月，果期10—11月。

**习性、应用及产地分布：** 温带及亚热带树种。喜光，不耐荫，较耐寒；萌蘖性强，病虫害少。树干端直，枝叶茂密，树形整齐优美，果形奇特，是良好的庭荫树及行道树，也可在草地、池畔等处孤植或丛植。中国特有种，栽培历史甚久，原产于中国中部及西部地区，四川、贵州、湖北等地为集中产区，现各地广泛栽培。

**Description:** Trees, deciduous, up to 20 m tall. Crown globose; bark dark gray, longitudinally fissured; branches, leaves, fruit and bark abundant in elastic glue. Young branches glabrate with flaky pith. Leaf blade ovate-oblong, 6–16 cm × 4–9 cm, apex acuminate, base rounded or broadly cuneate, margin serrate; reticulate veins prominent abaxially, pubescent. Samara narrowly oblong, 3–3.5 cm × 1–1.3 cm. Fl. Mar–May, fr. Oct–Nov.

# 17. 桑科

## 桑树 | *Morus alba*

**别名**：桑、白桑、家桑　　　　**科属**：桑科桑属

**形态特征**：落叶灌木或乔木，高 2～10 m。树冠倒卵圆形；树皮黄褐色，不规则浅纵裂；树体富含乳汁。叶卵形至广卵形，长 5～15 cm，宽 5～12 cm，先端渐尖或圆钝，基部圆形或浅心脏形，叶面光滑且有光泽，边缘有粗钝锯齿，幼树之叶常有不规则分裂。雌雄异株，花与叶同放；雄花序长 2～3.5 cm，下垂，密被白色柔毛；雌花序长 1～2 cm，无花柱。聚花果卵圆形或圆柱形，长 1～2.5 cm，成熟时黑紫色或红紫色。花期 4—5 月，果期 5—8 月。

**习性、应用及产地分布**：温带及亚热带树种。喜光，耐寒。生长快，萌芽力强，耐修剪；根系发达，抗风力强。树冠宽阔，树叶茂密，秋季叶色变黄，颇为美观；适生性强，易管理，为城市绿化先锋树种，宜孤植作庭荫树，或与其他树种混植为风景林。中国驯化栽培最早的树种之一，现南北各地广泛栽培，尤以长江中下游为多。

**Description:** Shrubs or trees to 2–10 m tall, deciduous. Crown obovate; bark yellowish brown, shallowly irregularly furrowed; abundant in latex. Leaf blade ovate to broadly ovate, 5–15 cm × 5–12 cm, apex acute or obtuse, base rounded to cordate, smooth and glossy, margin coarsely serrate to crenate. Dioecious, male catkins 2–3.5 cm, pubescent; female catkins 1–2 cm. Syncarp ovoid or cylindric, blackish purple or reddish purple when mature. Fl. Apr–May, fr. May–Aug.

# 构树 | *Broussonetia papyrifera*

**科属**：桑科构属

**形态特征**：落叶乔木，高16 m。树冠圆形或倒卵形；树皮灰褐色，光滑；枝条粗壮开展，小枝红褐色，密生白色绒毛。叶纸质，宽卵形或长卵形，长7～20 cm，宽6～15 cm，先端渐尖，基部略偏斜，上面被糙毛，下面密被柔毛，边缘有粗齿，不裂或不规则2～5深裂；三出脉；叶柄较长，被疏毛。雄花序长6～8 cm，雌花序球形；聚花果径1.5～2 cm，成熟橙红色。花期4—6月，果期8—9月。

**习性、应用及产地分布**：温带及亚热带树种。喜光，稍耐荫；对气候适应性强，耐干旱瘠薄，能生于水边。生长迅速，萌芽力强，根系较浅。对烟尘及有毒气体抗性很强。树形粗野，枝叶茂密，适应性强，是城乡、荒山坡地及严重污染的工矿区重要的绿化树种，可作庭荫树及防护林树种。分布广，中国辽宁南部、华北、西北、华中、华南、西南各地低山、平原均有分布。

**Description:** Trees, deciduous, 16 m tall. Crown globose or obovate; bark gray-brown, smooth; branches thick and strong. Leaf blade broadly ovate to narrowly elliptic-ovate, 7–20 cm × 6–15 cm, apex acuminate, base slightly oblique, margin coarsely serrate. Petiole long, sparsely pubescent, crude toothed edge, margin with no or 2-5 deep teeth irregularly, trinervious, long petiole, pubescent. Male inflorescences long spicate, 6–8 cm; female inflorescences globose. Syncarp 1.5–2 cm in diam., orange-red when mature. Fl. Apr–Jun, fr. Aug–Sep.

# 无花果 | *Ficus carica*

**别名：** 文仙果、仙人果　　　　　　　　**科属：** 桑科榕属

**形态特征：** 落叶灌木或小乔木，高 3～10 m。树皮暗褐色；多分枝，小枝直立，粗壮。叶大而厚，倒卵形或近圆形，长 10～15 cm，宽 8～14 cm，3～5 掌状深裂，先端钝，基部心形，上面粗糙，下面被短毛，边缘波状或具粗齿；托叶三角状卵形，淡红色。雌雄异株，隐头花序单生新枝叶腋。果实梨形或球形，顶部下陷，长 3～5 cm，径 3～4 cm，成熟时紫红色或黄色。花期 5—7 月，果期 10 月。

**习性、应用及产地分布：** 喜光，亦耐荫；喜温暖、稍干燥气候，不耐严寒，耐旱。生长快，根系发达。抗有毒气体能力较强，可在大气污染较严重的地区栽植。植于宅前屋后、园路旁、公园、街头绿地、池畔可点缀景色。原产于地中海沿岸，栽培历史悠久；中国南北均有零星栽培，以新疆南疆各地栽培最普遍，胶东及苏北盐碱土上生长良好。

**Description:** Shrubs, deciduous, 3–10 m tall; many branched. Bark grayish brown, distinctly lenticellate; branchlets straight, strong. Leaves alternate, petiole strong, leaf blade broadly ovate, 10–15 cm × 8–14 cm, usually with 3–5 ovate lobes, apex acuminate, base cordate, adaxially scabrous and short gray pubescence below, margin irregularly toothed. Dioecious, cryptic inflorescence, axillary on normal leafy shoots, solitary. Fruit pear-shaped, apical pore concave, 3–5 cm × 3–4 cm, purplish red to yellow when mature. Fl. May–Jul, fr. Oct.

# 天仙果 | *Ficus erecta*

**科属：** 桑科榕属

**形态特征：** 落叶小乔木，高2～7 m。树皮灰褐色。叶厚纸质，倒卵状椭圆形，长7～20 cm，宽3～9 cm，先端短渐尖，基部圆形至浅心形，全缘或上部偶有疏齿，叶面粗糙，两面疏生柔毛；侧脉5～7对。榕果单生叶腋，具总梗，球形或梨形，径约2 cm，幼时被柔短粗毛，顶生苞片脐状，成熟时黄红至紫黑色。花果期5—6月。

**习性、应用及产地分布：** 喜光，亦耐荫；喜温暖稍干燥气候。常植于宅前屋后、园路旁、林缘或溪边。产于中国广东、广西、贵州、湖北、湖南、江西、福建、浙江、台湾。生于山坡林下或溪边。

**Description:** Deciduous or semideciduous, 2–7 m tall. Bark grayish brown, branchlets glabrous or densely brown tomentose. Leaf blade, obovate-oblong, 7–20 cm × 3–9 cm, apex shortly acuminate or acute and mucronate, base rounded to cordate, margin entire or occasionally undulate toward apex, thickly papery, tomentose; 5–7 pairs of secondary veins. Fruit axillary on normal leafy shoots, solitary, peduncle, globose to pear-shaped, ca. 2 cm in diam., hairy when young, apical pore navel-like, reddish yellow to blackish purple or red when mature. Fl. and fr. May–Jun.

# 薜荔 | *Ficus pumila*

**科属：** 桑科榕属

**形态特征：** 常绿攀援藤本，幼时借气生根攀援。叶二型，营养枝上叶小而薄，卵状心形，长约 2.5 cm 或更短，先端渐尖，基部歪斜，近无柄；花序枝上的叶大而厚，革质，卵状椭圆形，长 5～10 cm，先端急尖至钝形，基部圆形至浅心形，叶下被黄褐色丝状柔毛，全缘；基出脉3条，网脉明显，侧脉3～4对，下面凸起成蜂窝状。花序托（隐花果）单生于叶腋，具短梗，果梨形或倒卵形，长3～6 cm，径约3 cm，熟时黄绿色微有红晕；基生苞片宿存。花期4—5月，果实9—10月成熟。

**习性、应用及产地分布：** 亚热带阴性树种。喜温暖、湿润气候，耐旱，适生于平原、丘陵和山麓腐殖质丰富的酸性土壤。性强健，生长快。叶质厚亮绿，寒冬不凋，宜攀援岩石壁、假山、墙垣、石桥等，可增强自然情趣；亦是优良的林下地被，可形成四季常春、花果并茂的观赏效果。产于中国河南、陕西、华东、华南至西南等地，日本和越南亦有分布。

**Description:** Evergreen lianas, climbers or scandent. Leaves distichous, leaf blade on fertile branchlets, 2.5 cm or shorter, apex acute, ovate-cordate, nearly no peduncle, veins conspicuous; secondary veins 3 or 4 on each side of midvein. Fruit axillary on normal leafy branches, solitary, short peduncle, pear-shaped to globose or cylindric, 3–6 cm × 3 cm, yellowish green to pale red when mature, involucral bracts triangular-ovate, persistent. Fl. Apr–May and fr. Aug–Sep.

# 柘树 | *Maclura tricuspidata*

**科属：** 桑科柘属

**形态特征：** 灌木或小乔木，高1～7 m。树皮淡灰色，成不规则薄片状剥落；小枝具有枝刺，刺长5～20 mm。叶卵形，长5～14 cm，宽3～6 cm，先端渐尖，基部近圆形，幼时两面有毛，老时仅叶背主脉被柔毛，全缘或偶为3裂；叶柄长1～2 cm。雄花序径约0.5 cm，雌花序径1～1.5 cm。果近球形，径约2.5 cm，肉质，成熟时橘红色。花期5—6月，果期9—10月。

**习性、应用及产地分布：** 温带阳性树种。耐寒，适生性强，耐干旱瘠薄，喜钙质土，生长速度缓慢。叶果秀丽，可在公园边角、街头绿地作庭荫树或刺篱，也是绿化荒滩、保持水土的先锋树种。产于中国华东、中南、西南、华北各省区，朝鲜、日本亦有分布。

**Description:** Shrubs or small trees, 1–7 m tall. Bark grayish brown, flaking in thin pieces irregularly; branchlets with 0.5–2 cm long spines. Leaves ovate, 5–14 cm × 3–6 cm, apex acuminate, base suborbicular, margin entire, occasionally 3-lobed, petiole 1–2 cm in length. Male inflorescences ca. 5 mm in diam., female inflorescences 1–1.5 cm in diam. Fruit orange red when mature, globose, ca. 2.5 cm in diam. Fl. May–Jun, fr. Sep–Oct.

## 18. 木兰科

### 广玉兰 | *Magnolia grandiflora*

**别名**：洋玉兰、荷花玉兰　　**科属**：木兰科木兰属

**形态特征**：常绿乔木，高达30 m。小枝、叶柄、叶背密被锈褐色短绒毛。叶厚革质，椭圆形或长圆状椭圆形，长10～20 cm，宽5～9 cm，先端钝圆，基部楔形，上面深绿色而有光泽，边缘略反卷；叶柄无托叶痕。花大，径15～20 cm，状如荷花，花被片9～12片，白色，芳香，厚肉质。聚合果圆柱形，长7～10 cm，密被灰褐色绒毛，蓇葖具长喙。花期5—6月，果期9—10月。

**习性、应用及产地分布**：亚热带阳性树种。喜光，适生于温暖、湿润气候，较耐寒。生长速度中等；根系深广，抗风能力强。病虫害少。树姿雄伟壮丽，叶大荫浓，花似荷花，芳香馥郁，为美丽的园林观赏树种，宜孤植、对植、丛植或列植为园景树、行道树、庭荫树。原产于北美东南部，中国长江流域以南广为引种栽培，已成为归化种。

**Description:** Trees, evergreen, up to 30 m tall. Twigs, petioles and abaxial leaf blade densely brownish tomentose. Leaves thickly leathery, elliptic or oblong-elliptic, 10−20 cm × 5−9 cm, apex obtuse, base cuneate, adaxially deep green and glossy. Flowers 15−20 cm in diam., lotus-like, tepals 12, white, fragrant, thickly fleshy. Fruit short terete, 7−10 cm long, densely grayish brown tomentose, follicle apex long beaked. Fl. May−Jun, fr. Sep−Oct.

## 玉兰 | *Magnolia denudata*

**别名：** 白玉兰、玉堂春　　　　　　**科属：** 木兰科木兰属

**形态特征：** 落叶乔木，高可达25 m。树皮灰白色，粗糙开裂；幼枝、冬芽及花梗密被灰黄色长绢毛。叶倒卵形或宽倒卵形，长10～18 cm，宽6～10 cm，先端平截或微凹，具突尖，基部楔形，叶背幼时疏被柔毛。花大，径12～15 cm，花被片9枚，白色、芳香，先叶开放；萼片与花瓣分化不明显。聚合果圆柱形，褐色，长10～13 cm，蓇葖木质，部分不发育而呈弯弓形。花期3月，果期8—9月。

**习性、应用及产地分布：** 喜光，亦能在半阴环境生长；对温度敏感，月平均气温5℃以上即可开花。肉质根，忌水涝，栽植地渍水易烂根。具一定的抗污染、滞尘和吸硫能力。中国著名的早春观花树种，花开时一片洁白，园林多植于庭园路边、草坪角隅、亭前院后或大型建筑物、纪念场所周围。原产于中国中部各省，现北京及黄河流域以南均有栽培。

**Description:** Trees, deciduous, up to 25 m tall. Bark grayish white, furrowed; twigs, winter buds, and pedicels densely grayish brown long tomentose. Leaves obovate or broad obovate, 10–18 cm × 6–10 cm, apex truncates, abruptly cuspidate, base cuneate. Flowers 12–15 cm in diam., tepals 9, white, fragrant, appearing before leaves. Fruit terete, 10–13 cm long, brown, follicle woody. Fl. Mar, fr. Aug–Sep.

## 紫玉兰 | *Magnolia liliflora*

**别名**：辛夷、木笔、木兰　　　　**科属**：木兰科木兰属

**形态特征**：落叶灌木，高4m。树皮灰褐色；小枝紫褐色，皮孔明显；冬芽被细毛。叶椭圆状倒卵形，长8～15cm，宽3～8cm，先端骤尖或渐尖，基部渐窄；侧脉8～10对。花蕾被淡黄色绢毛，酷似毛笔，花叶同放或花稍早于叶开放，花大，径10～15cm，花被片9～12片，外面紫色，里面近白色；萼片小，3枚，披针形；蓇葖果圆柱形，长7～10cm。花期3—4月，果期9—10月。

**习性、应用及产地分布**：暖温带至亚热带阳性树种。较耐寒，不耐旱；肉质根，忌积水。著名的庭园观赏树。早春开花，花时满树紫红，幽姿淑态，别具风情。用于古典园林中厅前院后配植，也可孤植或散植于小庭院内、窗前、假山石旁、池畔和水边。产于中国福建、湖北、四川、云南西北部及长江流域，南方各地广为栽培。

**Description:** Shrubs, deciduous, up to 4 m tall. Bark grayish brown; twigs purple brown, lenticels obvious, top buds tomentose. Leaves elliptically obovate, 8–15 cm × 3–8 cm, apex abruptly cuspidate or acuminate, base attenuate; secondary veins 8–10 pairs. Flower buds, yellowish tomentose, brush pen-like. Flowers big, 10–15 cm in diam., appearing with leaves or before leaves, petals 9–12, purple outside, white inside; sepals 3, small, lanceolate; follicle cylindric, 7–10 cm long. Fl. Mar–Apr, fr. Sep–Oct.

# 二乔玉兰 | *Magnolia soulangeana*

**别名：** 朱砂玉兰　　　　　　　　　　　**科属：** 木兰科木兰属

**形态特征：** 落叶小乔木，高可达10 m。叶革质，倒卵形，长10～15 cm，先端短急尖，基部楔形，叶背稍被柔毛；叶柄被柔毛。花大，钟状，径10～15 cm，先于叶开放；花被片6～9片，长圆状倒卵形，外面玫瑰红色，基部色较深，内面白色；萼片3片，花瓣状，长度达其半或与之等长。蓇葖果10～15 cm。花期2—3月，果期9—10月。

**习性、应用及产地分布：** 本种是玉兰与紫玉兰的杂交种。为著名的早春花木和园林观赏树种，花色绚丽多彩，耐寒性、耐旱性优于亲本。原产于中国，现在世界各地均有栽培。尤其在中国各地的庭园、公园和绿地中常有栽培。

**Description:** Small trees, deciduous, up to 10 m tall. Leaves leathery, obovate, 10–15 cm long, apex mucronate, base cuneate, abaxial surfaces sparsely tomentose. Flowers big, campanulate, 10–15 cm in diam., appearing before leaves, petals 6–9, oblong-obovate, rose-red outside, deep in base, white inside, sepals 3, petals-like, with equal or half length of the petals. Follicle ovate or obovate, 10–15 cm in diam. Fl. Feb–Mar, fr. Sep–Oct.

# 白兰 | *Michelia alba*

**别名：** 白兰花　　　　　　　　　　**科属：** 木兰科含笑属

**形态特征：** 常绿乔木，高可达18 m。叶薄革质，长椭圆形或披针状长圆形，长14～25 cm，宽5～9.5 cm，先端长渐尖，基部楔形，网脉两面明显；托叶痕达叶柄中部，叶柄长1.5～2 cm。花白色，极芳香，花被片10～14，披针形，长3～4 cm；雌蕊群被毛，心皮多数，部分心皮不发育。花期4—9月。

**习性、应用及产地分布：** 热带阳性树种。喜暖热、湿润气候，不耐低温。材质较脆，枝干易风折；忌烟尘。树形端直，花色洁白，夏秋开放，花香可持续数月，是传统的香花树种。作为中国华南地区的园林骨干树种，可群植或疏植于庭院、花坛、花境、草坪或道路两侧作园景树、行道树，用以营造优雅恬静的气氛。原产于印度尼西亚、斯里兰卡、印度等地，现广植于东南亚。

**Description:** Trees, evergreen, 18 m tall. Leaves thinly leathery, long elliptic to narrowly ovate, 10−25 cm × 5−9.5 cm, apex long acuminate, base cuneate, reticulate veins very conspicuous on both surfaces; stipular scar nearly reaching middle of petiole, petiole 1.5−2 cm. Flowers white, very fragrant, tepals 10−14, lanceolate, 3−4 cm long; gynoecium puberulous, carpels numerous, usually partly undeveloped. Fl. Apr−Sep.

## 含笑 | *Michelia figo*

**别名：** 香蕉花、含笑梅　　　　**科属：** 木兰科含笑属

**形态特征：** 常绿灌木，高 2～3 m。树冠圆形；分枝多；芽、嫩枝、叶柄、花梗均密被黄褐色绒毛。叶革质，窄椭圆形或倒卵状椭圆形，长 4～10 cm，宽 1.8～4 cm，先端短钝尖，基部楔形，全缘；叶柄长 2～4 mm，托叶痕达叶柄先端。花被片 6，肉质，淡乳黄色，边缘常带紫晕，肉质，具浓郁的香蕉香气；雌蕊群无毛。聚合果长 2～3 cm。花期 3—5 月，果期 7—8 月。

**习性、应用及产地分布：** 亚热带阴性树种。喜半阴环境，忌烈日曝晒；喜暖热湿润气候，有一定的耐寒性。是我国著名的香花树种，栽培历史悠久。常植于建筑物阴面、树下、疏林旁。原产于中国广东、福建一带的阴坡杂木林中，在长江流域及以南地区广为栽培。

**Description:** Shrubs, evergreen, 2–3 m tall. Crown globose; buds, young twigs, petioles, and pedicel densely yellowish brown tomentose. Leaves leathery, narrowly elliptic to obovate-elliptic, 4–10 cm × 1.8–4 cm, apex obtusely acute, base cuneate, margin entire. Petiole 2–4 mm, stipular scar reaching petiole apex. Tepals 6, fleshy, pale yellow, margin sometimes purple, sweetly fragrant. Gynophore glabrous. Fruit 2–3 cm. Fl. Mar–May, fr. Jul–Aug.

# 金叶含笑 | *Michelia foveolata*

**科属：** 木兰科含笑属

**形态特征：** 乔木，高可达30 m。幼枝及叶密被黄褐色绒毛，叶革质，长椭圆形至广披针形，长17～23 cm；叶面绿色，有光泽，叶背黄褐色绒毛。花被片9～12，白色，稍带黄绿色，基部带紫晕。花期3—4月。

**习性、应用及产地分布：** 喜温暖气候，较耐荫，生长较快。其嫩叶背面的金色绒毛在阳光下闪耀出金属光泽，有特殊的观赏价值。产于中国湖南、江西、福建、广东、云南等地。

**Description:** Trees to 30 m tall. Young twigs and leaf blade abaxial surfaces densely yellowish brown tomentulose. Leaf blade leathery, oblong-elliptic to broad lanceolate, 17–23 cm long, adaxially green and glossy, abaxial yellowish brown tomentose. Tepals 9–12, white, pale yellowish green, base purplish. Fl. Mar–Apr.

# 峨眉含笑 | *Michelia wilsonii*

**科属：** 木兰科含笑属

**形态特征：** 乔木，高达20 m。树皮光滑；叶革质，倒卵形至倒卵状披针形，长8～15 cm，叶背灰白色；网脉细密。花被片9（12），黄白色；雄蕊群细长，花径5～8 cm，芳香。聚合果成熟呈紫红色。花期3—5月。

**习性、应用及产地分布：** 树形优美，花大而洁白芳香，是良好的园林绿化及观赏树种。产于中国四川和湖北西部等地。

**Description:** Trees, up to 20 m tall. Bark smooth; leaf blade leathery, obovate to narrowly obovate, 8–15 cm long, abaxial grayish white; reticulate veins slender, dense. Tepals 9, yellow-white; stamens slender, flowers 5–6 cm in diam., fragrant. Fruit reddish purple when mature. Fl. Mar–May.

# 鹅掌楸 | *Liriodendron chinense*

**别名：** 马褂木、中国郁金香树　　　　**科属：** 木兰科鹅掌楸属

**形态特征：** 落叶乔木，高可达40 m。树皮灰白色；小枝灰褐色，具有环状托叶痕。单叶互生，有长柄，叶形似马褂，长6～14 cm，顶端平截或稍凹缺，两侧各具一深凹裂；叶背苍白色，密被白色乳头状凸起。花黄绿色，杯状，花被片9，长3～4 cm，稍直立，基部具黄绿色或深黄色条纹；雌蕊群超出花被片。聚合果纺锤状，长7～9 cm。花期5月，果期9—10月。

**习性、应用及产地分布：** 亚热带中性偏阴树种。喜凉润气候，在干旱、湿洼地生长不良。生长迅速，不耐移栽。世界著名的庭园观赏树。树干端直，绿荫如盖，分枝均匀，叶形奇特，花朵美丽。可孤植、列植为庭院、公园及广场庭荫树、行道树。中国特有种，产于陕西、安徽以南，西至四川、云南。国家二级重点保护植物。

**Description:** Trees, deciduous, up to 40 m tall. Bark grayish white; twigs grayish brown, with orbicular leaf scars. Leaf blade monophyllously alternate, with long petiole, T-shirt-like, 6–14 cm long, apex truncates or lightly emarginate, deeply lobed on both sides; abaxial pale white, with dense white papillae. Flower yellow-green, calicular, tepals 9, slightly upright, yellow-green or dark yellow stripes at base, gynoecium exceeding tepals. Aggregate fruit spindlelike, 7–9 cm long. Fl. May, fr. Sep–Oct.

## 北美鹅掌楸 | *Liriodendron tulipifera*

**别名**：郁金香树　　　　　　　　　　　　**科属**：木兰科鹅掌楸属

**形态特征**：落叶乔木，高可达60 m。小枝褐色或紫褐色，平滑。叶片先端近截形或微凹，两侧各具2～3个裂片，叶背无白粉。花被片9片，广卵形，长4～6 cm，浅黄绿色，基部具橙黄色蜜腺；雌蕊群超出花被片之上。聚合果纺锤状，长6～8 cm，翅状小坚果先端急尖。花期5月，果期9—10月。

**习性、应用及产地分布**：温带速生树种，喜光，耐寒，喜湿润。叶形奇特，花朵美丽，可孤植、列植于庭院观赏，或用作公园及广场庭荫树。原产于北美东南部。中国上海、南京、杭州、庐山、青岛、昆明等地有引种。

**Description:** Trees, deciduous, up to 60 m tall. Twigs brown or purplish brown, smooth. Leaf blade apex nearly truncate or lightly emarginate, with 2–3 shallow lobes on each side, abaxial surface without white farinose. Tepals 9, broad ovate, 4–6 cm long, light yellow-green, base with orange nectary; gynoecium not-exceeding tepals. Aggregate fruit spindlelike, 6–8 cm long, winglike small nut, apex abruptly acuminate. Fl. May, fr. Sep-Oct.

# 19. 蜡梅科

## 蜡梅 | *Chimonanthus praecox*

**别名**：黄梅、香梅    **科属**：蜡梅科蜡梅属

**形态特征**：落叶大灌木，高可达 4 m。根茎部发达呈疙瘩状，枝灰褐色，木质部芳香。叶对生，全缘，近革质，卵形或椭圆状披针形，先端渐尖，基部圆形或楔形，上面密被短糙毛。花单生于枝条两侧，先叶开放；花被片多数，黄色有光泽，似蜡质，具浓郁香味，内部具紫红色条纹或斑块；蒴果成熟时发育成坛状，口部收缩。花期 11 月下旬至翌年 3 月，果期 4—9 月。

**习性、应用及产地分布**：温带及亚热带树种。喜光、耐荫，较耐寒，耐旱性强。萌发力强，耐修剪，寿命长达百年。花凌寒怒放，花香四溢，是中国著名的冬春观赏花木，一般以自然式孤植、对植、丛植、群植等方式配置于园林与建筑物入口两侧和厅前亭周、窗前屋后、墙隅及草坪等处。产于中国秦岭、大巴山、武当山以南，现北京以南各地普遍栽培；日本、朝鲜自古从中国引种栽培，北美、欧洲、大洋洲亦有栽培。

**Description:** Shrubs or small trees, deciduous, 5 m tall. Stem developed in a pimple, branchlets grayish brown, wood aroma. Leaves opposite, entire, subleathery, leaf blade ovate, oblong-lanceolate, apex acuminate, base cuneate to rounded, adaxially roughly scabrous. Flowers on branches of previous year, solitary, appearing generally before leaves. Tepals 15–21, yellow, waxy, sweetly fragrant, inner ones usually with purplish red pigment. Capsule develops into an altar shape. Fl. Nov–Mar, fr. Apr–Sep.

# 亮叶蜡梅 | *Chimonanthus nitens*

**别名**：山蜡梅　　　　　　　　　　　**科属**：蜡梅科蜡梅属

**形态特征**：常绿灌木，高1～3 m。枝条被微毛，后渐无毛。叶小于蜡梅，卵状披针形，长5～13 cm，宽2～5.5 cm，先端长渐尖，基部楔形，叶面有光泽，略粗糙，叶背灰绿色，无毛或被白粉。花较小，径7～10 mm，窄尖，淡黄色或黄白色。果托坛状，长2～5 cm，口部缢缩，成熟时灰褐色，被短绒毛。花期10月至翌年1月，果期4—7月。

**习性、应用及产地分布**：耐荫，喜温暖、湿润气候及酸性土壤。根系发达，萌蘖力强。可引作观赏树。产于中国陕西秦岭南坡至长江流域，生于疏林或石灰岩山区。

**Description:** Shrubs, evergreen, 1–3 m tall. Branches slightly hairy, gradually hairless. Leaves opposite, ovate-lanceolate, 5–13 cm long, 2–5.5 cm wide, apex long acuminate, base cuneate, leaf blade lustrous, slightly rough, abaxially grayish green, sometimes with inconspicuous white powder. Flowers small, 7–10 mm in diam., narrow tip, pale yellow or yellow-white. Fruit altar-shaped, 2–5 cm long, gray-brown when mature, tomentose. Fl. Oct–Jan, fr. Apr–Jul.

# 20. 樟科

## 樟树 | *Cinnamomum camphora*

**别名**：香樟　　　　　　　　　　**科属**：樟科樟属

**形态特征**：常绿乔木，高 30～50 m。树皮黄褐色，纵裂；全株具樟脑香气；小枝亮绿色。叶互生，近革质，卵形或卵状椭圆形，长 7～12 cm，宽 3～5.5 cm，先端急尖，基部宽楔形至近圆形，下面微被白粉，边缘波状；离基三出脉，脉腋具腺体。圆锥花序，长 3.5～7 cm，总梗长 2.5～4.5 cm；花黄绿色，长约 3 mm。果球形，径约 6 mm，紫黑色，盘状果托肉质。花期 4—5 月，果期 8—11 月。

**习性、应用及产地分布**：喜光，喜温暖、湿润气候，耐水湿。萌芽力强，耐修剪，生长速度中等，寿命长；深根性，主根发达，抗风。吸烟滞尘，对多种有毒气体抗性较强。冠大荫浓，枝叶茂密，移栽易于成活，是江南最常见的园林绿化树种，适宜丛植、群植、孤植于池畔、水边、山坡、草坪，为庭荫树、行道树、背景树。主产于中国长江流域以南平原、丘陵地区。

**Description:** Trees, evergreen, 30–50 m tall. Bark yellow-brown, longitudinally fissured; branchlets shiny-green. Leaves alternate, subleathery, leaf blade ovate or ovate-elliptic, 7–12 cm × 3–5.5 cm, apex acute, base broadly cuneate or subrounded, abaxially slight white farinose, margin undulate, triplinerved. Panicle 3.5–7 cm, peduncle 2.5–4.5 cm. Flowers green-white, ca. 3 mm. Fruit globose, ca. 6 mm in diam., purple-black. Fl. Apr–May, fr. Aug–Nov.

# 天竺桂 | *Cinnamomum japonicum*

**别名：**竺香、山肉桂　　　　　　　　　**科属：**樟科樟属

**形态特征：**常绿乔木，高10～15 m。枝条圆柱状，红色或红褐色，纤细。叶互生或近对生，革质，卵圆状长圆形至长圆状披针状，长7～10 cm，宽3～3.5 cm，先端锐尖至渐尖，基部阔楔形至钝形，下面灰绿色，全缘；离基三出脉近平行，并在叶两面隆起。腋生圆锥花序长，花被片内外被柔毛，花被筒倒锥形，长约1.5 mm。果长圆形，长约7 mm，果托浅杯状。花期4—5月，果期7—9月。

**习性、应用及产地分布：**亚热带中性树种。喜温暖、湿润气候。树形端正优美，枝叶茂密，四季翠绿，在园林中孤植、丛植、列植均可。产于中国华东、华中、华南至台湾。

**Description:** Trees, evergreen, 10 ～ 15 m tall. Branches terete, red or reddish-brown, slender. Leaves alternate or nearly opposite, leathery, ovate-oblong to oblong-lanceolate, 7–10 cm × 3–3.5 cm, apex acute to acuminate, base broadly cuneate to obtuse, abaxially gray-green, entire; triplinerved, nearly parallelveined, veins impressed on both sides. Axillary panicles long, pilose tomentose, perianth tube inverted, 1.5 mm long. Fruit oblong, 7 mm long, penianth cup in fruit shallowly cupuliform. Fl. Apr–May, fr. Jul–Sep.

# 浙江楠 | *Phoebe chekiangensis*

**别名：** 浙江紫楠　　　　　　　　　**科属：** 樟科楠属

**形态特征：** 常绿乔木，高可达20 m。树皮淡黄褐色，薄片脱落；小枝密被黄褐色绒毛。单叶互生，全缘，倒卵状椭圆形至倒卵状披针形，长8～13 cm，宽3.5～5 cm，先端渐尖，基部楔形或近圆，叶缘向后反卷，叶幼时被毛，后渐无毛，叶背密被灰褐色柔毛，中脉在叶背明显。圆锥花序，长5～10 cm，被毛；花被片被毛。果卵圆形，长1.2～1.5 cm，被白粉。花期4—5月，果期9—10月。

**习性、应用及产地分布：** 耐荫树种，具有深根性，抗风强。浙江楠树体高大通直，端庄美观，具有较高的观赏价值。宜作庭荫树、行道树或风景树，或在草坪中孤植、丛植，也可在大型建筑物前后配置。分布于中国浙江、福建北部及江西东部。

**Description:** Trees, evergreen, up to 20 m tall. Bark pale yellowish brown; branchlets densely yellowish brown tomentose. Leaves alternate, margin entire and slightly recurved, obovate-elliptic to obovate-lanceolate, 8–13 cm × 3.5–5 cm, apex acuminate, base cuneate or subrounded, abaxially grayish brown pubescent, midrib adaxially impressed. Panicles 5–10 cm, tomentose; perianth tomentose. Fruit ellipsoid-ovoid, 1.2–1.5 cm long, with white farinose. Fl. Apr–May, fr. Sep–Oct.

# 红楠 | *Machilus thunbergii*

**别名：** 红润楠　　　　　　　　　　**科属：** 樟科润楠属

**形态特征：** 常绿乔木，高20 m。树皮黄褐色，幼枝带紫红色。单叶互生，常集生枝顶，革质，全缘，倒卵形，长4.5～9 cm，先端短突尖，基部楔形，叶面有光泽，叶背被白粉；羽状侧脉7～12对；叶柄带红色，长1～3.5 cm。圆锥花序顶生或生于叶腋，总梗紫红色。果扁球形，径约1 cm，熟时紫黑色；基部反卷花被片宿存，果序梗及果梗增粗，肉质鲜红色。花期3—4月，果期6—7月。

**习性、应用及产地分布：** 喜光，稍耐荫，为同属树种中较耐寒者；喜湿，具较强的耐盐及抗风能力；萌芽性强。树形优美，枝叶浓密，宜植作庭荫树、行道树及风景树，东南沿海地区可作防风林。产于中国福建、两广、江西、湖南、湖北和台湾等地。

**Description:** Trees, evergreen, up to 20 m tall. Bark yellowish brown, young branchlets purple-brown. Leaves alternate, leathery, margin entire, obovate, 4.5–9 cm long, apex abruptly cuspidate, base cuneate, leaf blade lustrous, abaxially with white farinose; lateral veins 7–12 pairs; petiole reddish, 1–3.5 cm long. Panicle axillary or subterminal, peduncle reddish purple. Fruit compressed globose, ca. 1 cm in diam., becoming dark purple when mature; perianth lobes persistent, fruiting pedicels thickened, fleshy, reddish. Fl. Mar–Apr, fr. Jun–Jul.

# 薄叶润楠 | *Machilus leptophylla*

**别名**：华东楠　　　　　　　　　**科属**：樟科润楠属

**形态特征**：常绿大乔木，高可达28 m。树皮灰褐色。叶互生，常集生于枝端，倒卵状长圆形，长14～25 cm，先端短渐尖，基部楔形，背面白粉显著；中脉在叶面凹下，在背面显著凸起，侧脉14～21对。圆锥花序，花白色。果球形，径约1 cm。

**习性、应用及产地分布**：稍耐荫，喜温暖、湿润气候，生长较快。树大荫浓，树形美观，可作庭荫树。产于中国江苏、安徽、福建、浙江、江西、湖南、两广及贵州等地。

**Description:** Large trees, evergreen, up to 28 m tall. Bark grayish brown. Leaves alternate, grows on apical part of branchlets, obovate-oblong, 14−25 cm long, apex shortly acuminate, base cuneate, abaxially with distinct white farinose; midrib impressed abaxially, lateral veins 14−21 pairs. Panicles, flowers white. Fruit globose, ca. 1 cm in diam.

# 刨花楠 | *Machilus pauhoi*

**别名：**刨花润楠　　　　　　　　　　　　**科属：**樟科润楠属

**形态特征：**常绿乔木，高达20 m，树皮灰褐色，有浅裂；小枝无毛。叶常集生于枝端，披针形至倒披针形，长6～15 cm，先端渐尖，基部楔形，背部有白粉；中脉下凹，在叶背明显突起，侧脉8～14对，叶干后黑色。圆锥花序生于新枝下部。果球形，径约1 cm，熟时黑色。

**习性、应用及产地分布：**喜温暖气候及排水良好土壤，萌芽力强，深根性，抗风。树姿雄伟，无病虫害，是优良的园林绿化及观赏树种。产于中国东南至华南山地。

**Description:** Trees, evergreen, up to 20 m tall. Bark grayish brown, slightly fissured; branchlets glabrous. Leaves grow on apical part of branchlets, lanceolate to oblanceolate, 6–15 cm long, apex acuminate, base cuneate, abaxially with white farinose, concave adaxially; midrib distinctly elevated abaxially, leaves black when drying, lateral veins 8–14 pairs. Cymose panicles on lower part of 1-year-old branch. Fruit globose, ca. 1 cm in diam., black when mature.

# 山胡椒 | *Lindera glauca*

**科属**：樟科山胡椒属

**形态特征**：落叶灌木或小乔木，高可达8 m。小枝灰白色，幼时有毛。单叶互生，近纸质，卵形或卵状椭圆形，长4～9 cm，羽状脉，全缘，叶背苍白色，具灰色柔毛。果球形，径约7 mm，成熟时黑色。

**习性、应用及产地分布**：喜光，耐干旱瘠薄；深根性。叶片芳香，秋叶红色，可作庭园观赏。产于中国黄河以南地区。

**Description:** Shrubs or trees, deciduous, up to 8 m tall. Branchs gray-white, pubescent when young. Leaves alternate, subpapery, leaf blade ovate or ovate-ellipsoid, 4–9 cm long, pinninerved, margin entire, white abaxially with gray pubescent. Fruit globose, ca. 7 mm in diam., black when mature.

## 狭叶山胡椒 | *Lindera angustifolia*

**科属：** 樟科山胡椒属

**形态特征：** 落叶灌木，高2～8 m。小枝黄绿色，无毛。叶长椭圆状披针形，长5～14 cm，羽状脉，全缘，叶背疏生细长毛；网状脉隆起。花芽生于叶芽两侧，伞形花序无总梗。核果球形，直径约8 mm，黑色，无毛。花期3—4月，果期9—10月。

**习性、应用及产地分布：** 喜光，耐干旱贫瘠。分布于中国华东、华中及华南。

**Description:** Deciduous shrubs, up to 2-8 m tall. Young branchlets yellow-green, glabrous. Leaf blade elliptic-lanceolate, 5-14 cm long, pinninerved, margin entire, pale and laxly pubescent abaxially; veins distinctly elevated. Flowers buds born on sides of leaf buds, umbels without peduncle. Fruit drupaceous, globose, ca. 8 mm in diam., black, glabrous. Fl. Mar-Apr, fr. Sep-Oct.

# 月桂 | *Laurus nobilis*

**科属：** 樟科月桂属

**形态特征：** 常绿小乔木，高 10～12 m。树冠卵形；树皮黑褐色；小枝绿色，全株具香气。叶互生，硬革质，长椭圆形至广披针形，长 6～12 cm，羽状脉，先端渐尖，基部楔形，叶面暗绿色，有光泽，叶背淡绿色，边缘细波状；叶柄带紫色。花小而不显，黄色，有花 5 成紧密聚伞状花序簇生叶腋，总梗长 5～10 mm。核果椭圆形，成熟时黑色或暗紫色。花期 3—5 月，果期 6—10 月。

**习性、应用及产地分布：** 喜光，稍耐荫；喜温暖、湿润气候，耐干旱，怕水涝。生长速度较快，萌芽力强。树形圆整，枝叶茂密，四季常青，是良好的庭园绿化树种。孤植、丛植于草坪，还可修剪成球体、长方体等用于草地、公园、街头绿地的点缀。原产于地中海沿岸，中国江苏、浙江、福建、台湾、四川及云南等地均有引种栽培。

**Description:** Evergreen trees, up to 10−12 m tall. Crown ovate; bark dark brown, branchlets green when young, aroma. Leaves alternate, leathery, oblong to broadly lanceolate, 6−12 cm long, pinnate veins, apex acuminate, base cuneate, leaf blade dark green, lustrous, abaxially light green, margin fine wave; petiole purple. Flowers small and not obvious, yellow, flowers 5 dense cymose fascicled, axillary, peduncle 5−10 mm long. Drupe elliptic, black or dark purple when mature. Fl. Mar−May, fr. Jun−Oct.

## 舟山新木姜子 | *Neolitsea sericea*

**科属：** 樟科新木姜子属

**形态特征：** 乔木，高 10 m。树皮灰白色、平滑；嫩枝密被金黄色丝状柔毛，老枝紫褐色。叶革质，长椭圆形或卵状长椭圆形，长 6.6～20 cm，宽 3～4.5 cm，幼叶两面密被金黄色绢状毛，老叶上面深绿色且有光泽，下面粉绿色，密被毛，边缘反卷；离基三出脉；叶柄长 2～3 cm。花 5 朵成伞形花序簇生于枝端叶腋；雌花退化，雌蕊被毛。核果椭圆形，径 8～10 mm，成熟时鲜红色，果托浅杯状。花期 9—10 月，果期翌年 1—2 月。

**习性、应用及产地分布：** 耐荫，喜温暖、湿度高的海岛生境；耐旱，生长迅速，根系发达，萌蘖力较强，抗风。枝叶繁茂，四季常青，冬季红果满枝，与绿叶相映，十分艳丽，是中国长江流域及东南沿海城市不可多得的优良观赏树种，适宜于庭院及"四旁"（指宅旁、村旁、路旁、水旁）绿化。国家二级重点保护植物。

**Description:** Trees to 10 m tall. Young branchlets and petioles with dense golden yellow sericeous pubescence. Leaves alternate, 6.6–20 cm × 3–4.5 cm, leaf blade elliptic to lanceolate-elliptic, with dense golden yellow sericeous tomentose on both surfaces, shiny adaxially. Umbels, axillary or lateral, sessile, clustered, 5-flowered. Fruit globose, ca. 1 cm in diam., seated on shallowly discoid perianth tube. Fl. Sep–Oct, fr. Jan–Feb.

# 21. 小檗科

## 豪猪刺 | *Berberis julianae*

**科属：** 小檗科小檗属

**形态特征：** 常绿灌木，分枝紧密，高2～2.5 m。小枝黄褐色，有棱角；有三叉刺，刺长达3.5 cm。叶狭椭圆形至倒披针形，长5～7.5 cm，宽0.8～1.5 cm，叶缘有刺齿6～10对。花常15～20朵簇生于叶腋，黄色，微香，径约6 mm，有细长柄；浆果卵形，成熟时蓝黑色，被白粉。花期5—6月，果期秋季。

**习性、应用及产地分布：** 性较耐寒。适于庭园观赏。产于中国中部地区。

**Description:** Shrubs, evergreen, 2–2.5 m tall. Branches yellow-brown, sulcate; spines 3-fid, 3.5 cm long. Leaf blade narrow-elliptic or oblanceolate, 5–7.5 cm × 0.8–1.5 cm, margin 6–10 spinose-serrate on each side. Flowers 10–25 fascicled axillary, yellow, slightly fragrant, ca. 6 mm in diam., pedicels slender. Berry ovate, blue-black when mature, white pruinose. Fl. May–Jun, fr. autumn.

# 阔叶十大功劳 | *Mahonia bealei*

**别名：** 刺黄檗、黄天竹  **科属：** 小檗科十大功劳属

**形态特征：** 小灌木，高0.5～4 m。羽状复叶长27～51 cm，具小叶7～15枚，硬革质，叶面绿色有光泽，叶下被白粉，边缘两侧具2～5对粗刺齿；顶生小叶近圆形，侧生小叶基部偏斜，无叶柄。3～9个直立的总状花序簇生；花黄色，花梗长4～6 cm；花瓣倒卵状椭圆形，长6～7 mm，宽3～4 mm，基部腺体明显，先端微凹。浆果卵形，长1.5 cm，熟时深蓝色，被白粉。花期9月至翌年1月，果期3—5月。

**习性、应用及产地分布：** 温带阴性树种。喜温暖气候，不耐严寒，萌蘖性较强。四季常绿，叶形奇特，常与山石配植用于布置庭院、花坛、岩石园、水榭，也适于建筑物附近的绿化。产于中国黄河以南各省。

**Description:** Small trees to 0.5–4 m tall. Pinnately compound leaves, 27–51 cm, with 7–15 pairs of leaflets, rigid-leathery, adaxially green, lustrous, abaxially with white farinose, margin with 2–5 teeth on each side. Inflorescence erect, 3–9-fascicled racemes; flowers yellow, pedicel 4–6 mm; petals obovate-elliptic, 6–7 mm × 3–4 mm, base with distinct glands, apex slightly emarginate. Berry ovoid, 1.5 cm long, dark blue when mature, pruinose. Fl. Sep–Jan, fr. Mar–May.

## 狭叶十大功劳 | *Mahonia fortunei*

**别名**：猫儿头、细叶十大功劳　　　　**科属**：小檗科十大功劳属

**形态特征**：常绿小灌木，高0.5～2 m。叶革质，羽状复叶具小叶5～9枚，小叶狭披针形，叶面暗绿色至深绿色，有光泽，边缘具5～10对刺齿；叶脉不显。4～10个总状花序簇生，长3～7 cm，花梗与苞片等长；苞片卵形；花瓣6枚，基部腺体显著。浆果球形，径4～6 mm，熟时紫黑色，被白粉。花期7—9月，果期9—11月。

**习性、应用及产地分布**：亚热带阴性树种。喜温暖、湿润气候，耐寒性差，较耐旱，忌水涝；萌蘖力强。叶形奇特，花朵鲜黄色，在长江流域常植于庭园、林缘及草地边缘，也可用于点缀假山、岩石或作绿篱材料。产于中国长江以南地区。

**Description:** Shrubs, evergreen, 0.5–2 m tall. Pinnately compound leaves, leaflets 5–9, leathery, narrowly lanceolate, adaxially dark green to deep green, glossy; veins inconspicuous. Margin with 5–10 spinose teeth on each side. Inflorescence erect, 4–10 fascicled racemes, floral bracts ovate, petals 6, base with distinct glands. Berry globose, 4–6 mm in diam., purple-black when mature, pruinose. Fl. Jul–Sep, fr. Sep–Nov.

# 南天竹 | *Nandina domestica*

**别名：** 蓝田竹、天竺　　　　　　　　　　　**科属：** 小檗科南天竹属

**形态特征：** 丛生灌木，高 3 m。2～3 回羽状复叶集生于茎上部，长 30～50 cm；小叶薄革质，椭圆形或椭圆状披针形，叶面深绿色，冬季常变红色。顶生圆锥花序，长 20～35 cm，小花白色，芳香；萼片多轮，花瓣长圆形，先端圆钝。浆果球形，径 5～8 mm，熟时鲜红色，经冬不落。花期 3—6 月，果期 9—10 月。

**习性、应用及产地分布：** 亚热带树种。喜半荫，适生于温暖气候，耐寒性不强。生长速度较慢，萌蘖力强，寿命长。枝干挺拔如竹，羽叶秀美，春赏绿叶，夏观白花，秋冬红果累累，色泽艳丽，是观叶赏果之佳品。园林中常丛植于庭院、山石、花台、草地边缘、路旁、水边或园路转角处，红绿相映，景色宜人。中国长江流域及其以南地区庭园多栽培。

**Description:** Shrubs to 3 m tall. Leaves glabrous, blades elliptic or elliptic-lanceolate, adaxially deep green, turning red in winter. Panicle, terminal, 20–35 cm long. Flowers white, fragrant; sepals multiple whorls, petals oblong, apex obtuse. Berry globose, 5–8 mm in diam., red when mature. Fl. Mar–Jun, fr. Sep–Oct.

## 22. 猕猴桃科

### 中华猕猴桃 | *Actinidia chinensis*

**别名：** 猕猴桃、羊桃藤　　　　　　**科属：** 猕猴桃科猕猴桃属

**形态特征：** 落叶木质藤本。小枝柔细，具皮孔，髓大，片层状。叶圆形或卵圆形，长6～8 cm，宽7～8 cm，先端多微凹或平截，基部钝圆或浅心形，上面仅脉上被疏毛，下面密生灰白色或淡褐色星状绒毛，边缘具睫状细齿；侧脉5～8对；叶柄长3～6 cm，被毛。花单生或3朵成聚伞花序；花初开时白色，后变淡黄色，径约2.5 cm，芳香；萼片密被黄褐色绒毛；花瓣3～7枚；雄蕊极多；子房球形，密被金黄色绒毛。浆果近球形或椭圆形，长4～4.5 cm，被柔毛。花期5—6月，果期9—10月。

**习性、应用及产地分布：** 亚热带或温带树种。喜光，略耐荫；喜温暖，有一定耐寒能力。不耐涝、不耐旱。萌芽力强，耐修剪。藤蔓虬攀，花色艳丽，芳香；果实圆大，可用于绿化廊架、篱垣、园门，攀附于山石、陡壁。中国特有藤本树，广泛分布于长江流域以南。

**Description:** Climbing lianas, deciduous. Branchlets slender, with lenticels, pith lamellate. Leaf blade orbicular or broadly ovate, 6–8 cm × 7–8 cm, apex emarginate or truncate, base rounded or cordiform, adaxially glabrous; only densely puberulent on midvein and lateral veins, abaxially gray-white or pale brown stellate tomentose. Ovary globose, densely golden villous. Berry subglobose or ellipsoid, 4–4.5 cm, tomentose. Fl. May–Jun, fr. Sep–Oct.

# 23. 山茶科

## 山茶 | *Camellia japonica*

**别名：** 茶花、曼佗罗、野山茶  **科属：** 山茶科山茶属

**形态特征：** 常绿小乔木，高 6～9 m。树皮灰白色，平滑；冬芽无毛，嫩枝淡褐色。叶革质，卵形或椭圆形，长 5～10 cm，宽 2.5～5 cm，叶面深绿色，且具光泽、具钝齿；侧脉不明显；叶柄长 8～15 mm。花单生或 2～3 朵对生于叶腋或枝顶，花梗极短；花瓣 5～7 枚，或为重瓣，基部明显合生，通常红色，径 6～10 cm；子房无毛，3～4 室，花柱 3～5 裂。蒴果圆球形，径 2.5～3 cm。花期 11 月至翌年 4 月，果期翌年 9—10 月。

**习性、应用及产地分布：** 亚热带树种。喜温暖、湿润气候。中国"十大"传统名花之一。树形多姿，叶色翠绿，花大艳丽，花期长；常丛植或散植于庭园、花境、假山旁或草坪、树丛边缘用于装点景色，也可片植或林植为专类园。中国是山茶属原始物种起源和分布中心，长江流域及其以南各省可露地栽培。

**Description:** Small evergreen trees, 6–9 m tall. Bark gray-white, smooth; winter buds glabrous, young branches grayish brown. Leaf blade leathery, ovate or elliptic, 5–10 cm × 2.5–5 cm, adaxially dark green and lustrous, margin serrulate, petiole 8–15 mm long. Flowers axillary or subterminal, solitary or in clusters of 2–3, subsessile, petals 5–7 or double, basally connate, 6–10 cm in diam. Ovary glabrous, 3–4 loculed, apically 3–5 lobed. Capsule globose, 2.5–3 cm in diam. Fl. Nov–Apr of following year, fr. Sep–Oct.

# 油茶 | *Camellia oleifera*

**别名：** 茶子树、白花茶　　　　　　　　**科属：** 山茶科山茶属

**形态特征：** 常绿小乔木，高可达7 m。树皮淡黄褐色，平滑不裂。冬芽密被金黄色长柔毛。叶厚革质，椭圆形或倒卵形，长5～7 cm，宽2～4 cm，先端尖，基部楔形；叶面有光泽，中脉显著且被毛，边缘具细锯齿；叶柄长4～8 mm，被毛。花1～2朵顶生，径3～8 cm，白色，无花梗；花瓣5～7枚，先端凹；子房被白色丝毛，3室。蒴果球形，径2～4 cm。花期9月至翌年2月，果期翌年秋季。

**习性、应用及产地分布：** 阳性树种。适生于温暖、湿润气候，不耐盐碱。深根性，萌蘖性较强；生长缓慢，寿命长。观赏兼经济树种。枝叶茂密，盛花期满树银花，素雅馥郁，可丛植为花篱或于林缘配植，在大面积风景林中宜作背景树种。产于中国秦岭、淮河以南，印度、越南亦有分布。

**Description:** Small evergreen trees, 7 m tall. Bark pale yellowish brown, smooth. Winter buds densely golden villous. Leaf blade thickly leathery or obovate, 5–7 cm × 2–4 cm, apex acute, base cuneate, margin serrate to serrulate, adaxially lustrous; midvein impressed and tomentose, petiole 4–8 mm, tomentose. Flowers 3–8 cm in diam., white, subsessile, petals 5–7, apex emarginate, ovary white tomentose, 3-loculed. Capsule globose, 2–4 cm in diam. Fl. Sep–Feb of following year, fr. autumn of following year.

# 茶梅 | *Camellia sasanqua*

**别名：** 茶梅花　　　　　　　　　　　　**科属：** 山茶科山茶属

**形态特征：** 常绿灌木或小乔木，高3～6 m。树皮灰白色，幼枝被粗毛。叶较小而厚，革质，椭圆形至长圆卵形，长4～8 cm，宽2～3 cm；叶面具光泽，叶背褐绿色，边缘有细锯齿；网脉不显著。花顶生，白色或红色，略芳香，大小不一，径4～7 cm，无花梗；花瓣6～7枚，阔倒卵形，近离生；雄蕊离生；子房被绒毛。蒴果球形，径1.5～2 cm，稍被毛。花期10月至翌年4月；果期翌年8—9月。

**习性、应用及产地分布：** 亚热带树种。性强健，适生于温暖、阴湿环境，耐修剪。树形优美、花叶茂盛品种可孤植或对植于庭院和草坪中；较低矮者可与其他花灌木配置点缀花坛、花境、林缘、墙基等处，亦可作为花篱。主产于中国江苏、浙江、福建、广东等南方各省，日本亦有分布。

**Description:** Evergreen shrubs or small trees, 3 to 6 m tall. The bark is grayish white and the young branches are rough. The leaves are small and thick, leathery, elliptic to oblong-ovate, 4–8 cm × 2–3 cm, adaxially glossy, abaxially brownish green, margins finely serrated. Veins are not significant. Flowers solitary, terminal, white or red, slightly aromatic, varying in size from 4 to 7 cm in diam., sessile; petals 6–7, broadly obovate; stamens separate, ovate tomentose. Capsule globose, 1.5–2 cm in diam. Fl. Oct–Apr of following year, fr. Aug–Sep.

# 杜鹃红山茶 | *Camellia azalea*

**别名**：杜鹃茶、四季茶　　　　　　　　**科属**：山茶科山茶属

**形态特征**：常绿灌木，高1～2.5 m。花似山茶，叶像杜鹃，株型紧凑，分枝密；老枝光滑，灰褐色；嫩枝无毛，略显红色。叶长倒卵形或倒披针形，先端圆钝或微凹，基部楔形；叶两面均无毛，稍被灰粉；叶全缘，边缘具透明骨质状狭边，叶多聚集于枝梢上部。花常生于枝端，红色或粉色，无花梗；花瓣5～9枚。子房卵形无毛，蒴果卵圆形或纺锤形，成熟时果皮由青色变成褐色。四季开花不断，盛花期7—9月。

**习性、应用及产地分布**：半阳性树种。具有较高的观赏价值，一度曾濒临灭绝。杜鹃红山茶原产于中国，其分布地区极窄，原只在广东省阳春市境内有零星分布，野生数量稀少，国家一级保护植物。

**Description:** Shrubs, evergreen, 1–2.5 m tall. Old branches smooth, grayish brown; young branches glabrous, red. Leaf blade oblong-obovate or oblanceolate, apex obtuse or with an obtuse tip, base cuneate; both surfaces glabrous, margin entire. Flowers always subterminal, red or pink, subsessile, petals 5–9. Ovary ovoid, glabrous. Capsule ovoid or fusiform, seedcase green turning brown when mature. Fl. throughout the year.

## 厚皮香 | *Ternstroemia gymnanthera*

**科属：** 山茶科厚皮香属

**形态特征：** 常绿灌木至小乔木，高3～8m。树皮灰褐色，平滑；小枝粗壮。叶革质，长椭圆状倒卵形，长5.5～9cm，宽2～3.5cm，先端锐尖，基部楔形，叶面光亮，全缘；侧脉不显；叶柄长1～1.3cm。花单生于当年生无叶小枝叶腋，径1～1.4cm，花梗长约1cm；萼片5枚；花瓣倒卵形，淡黄白色；雄蕊约50枚，基部与花瓣连生；子房2室，每室胚珠2枚。浆果状蒴果球形，径7～10mm，熟时紫红色；每室种子1粒，肾形，肉质假种皮红色。花期5—7月，果期8—10月。

**习性、应用及产地分布：** 亚热带阴性树种。喜温热、湿润气候，能耐-10℃低温。适生于肥沃的酸性或中性土壤。根系发达，抗风，移栽易成活；萌芽力弱，不耐强度修剪。树冠浓绿，花香果红，冬叶绯红，分外艳丽。大树适宜植于大门两侧、步道及草坪边缘，幼树可植作绿篱或散植、丛植用于配置成景。主产于中国华中、华南、西南至台湾各省。

**Description:** Shrubs, evergreen, 3–8 m tall. Bark grayish brown, smooth; young branches stout. Leaf blade leathery oblong-obovate, 5.5–9 cm × 2–3.5 cm, apex shortly acuminate, base cuneate, adaxially shiny, margin entire, secondary veins obscure. Petiole 1–1.3 cm. Flowers solitary, axillary on 1st-year branchlets, 1–1.4 cm in diam., sepals 5, petals obovate, pale yellow-white. Fruit globose, 7–10 mm in diam., purplish red when mature. Fl. May–Jul, fr. Aug–Oct.

# 24. 藤黄科

## 金丝桃 | *Hypericum monogynum*

**别名：** 金丝海棠　　　　　　　　　　**科属：** 藤黄科金丝桃属

**形态特征：** 半常绿灌木，高0.6～1 m。多分枝，小枝红褐色，光滑无毛。叶对生，长椭圆形，长3～10 cm，宽2～3 cm，稍抱茎，叶背粉绿色，全缘，无叶柄。花单生或成顶生聚伞花序，径4～5 cm；花瓣宽倒卵形，金黄色；雄蕊花丝较花瓣长；花柱细长，先端5裂。蒴果卵圆形，长约1 cm。花期6—7月，果期8—9月。

**习性、应用及产地分布：** 温带及亚热带阳性树种。不耐寒，忌积水。根系发达，萌芽力强，耐修剪。花丝纤细，灿若金丝，花冠似桃，故名"金丝桃"。仲夏黄花密集，为南方夏季庭院中常见的观赏花木，常植作绿篱、花境，也可丛植、群植于园林景观中。主要产于中国黄河以南地区，日本亦有分布。

**Description:** Shrubs, semi-evergreen, 0.6–1 m tall. Branches bushy, young branches reddish brown, smooth, glabrous. Leaves opposite, blade oblong, 3–10 cm × 2–3 cm, abaxially pink-green, margin entire, petiole absent. Inflorescence solitary or cymose subterminal, 4–5 cm in diam. Petals broadly obovate, golden yellow, stamen longer than petals. Capsule ovoid, ca. 1 cm long. Fl. Jun–Jul, fr. Aug–Sep.

## 金丝梅 | *Hypericum patulum*

**别名**：遍地金、云南连翘　　　　　　**科属**：藤黄科金丝桃属

**形态特征**：半常绿灌木，高 0.3～1.5 m。小枝拱曲，红色或暗褐色。叶较小，对生，卵状长椭圆形或广披针形，长 2～5 cm，宽 1.5～3 cm，叶背散布油腺点；叶柄极短。花金黄色，径 4～5 cm；花丝短于花瓣，不伸出花冠外；花柱 5 枚，离生。蒴果卵形，长约 1 cm，具宿萼。花期 4—8 月，果期 6—10 月。

**习性、应用及产地分布**：亚热带及热带阳性树种。有一定耐寒力，不耐积水。萌芽力强，耐修剪。园林应用同金丝桃。枝叶丰满，色彩鲜艳，可丛植或群植于草坪、树坛边缘和墙角、路旁等处。产于中国华中、东南和西南地区，日本、非洲南部已归化，其他各国常见栽培。

**Description:** Shrubs, semi-evergreen, 0.3–1.5 m tall. Branchlets arching, red or dark brown. Leaves opposite, blade oblong-ovate or broadly lanceolate, 2–5 cm × 1.5–3 cm, abaxial glands dense, petiole short. Flowers golden yellow, 4–5 cm in diam. Filaments shorter than petals, not spreading out of corolla, styles 5. Capsule ovoid, ca. 1 cm long, sepals persistant. Fl. Apr–Aug, fr. Jun–Oct.

## 25. 连香树科

### 连香树 | *Cercidiphyllum japonicum*

**科属：** 连香树科连香树属

**形态特征：** 落叶乔木，高可达10～20 m。树皮棕灰色；单叶对生，广卵圆形，基部心形，长4～7 cm，宽3.5～6 cm，掌状脉5～7条，边缘有圆钝锯齿，叶背面灰绿色带粉霜。雄花常4朵丛生，近无梗；雌花2～6朵丛生；蓇葖果2～4个，荚果状。花期4月，果期8月。

**习性、应用及产地分布：** 喜光，喜温凉、湿润气候及肥沃土壤，萌蘖性强。形优雅，幼叶紫色，秋叶黄、红或紫色，是优美的山林风景树及庭荫、观赏树种。产于中国中西部山地及日本，为古老的孑遗树种。

**Description:** Trees, deciduous, up to 10–20 m tall. Bark brownish gray; leaves opposite, broadly ovate, base heart-shaped, 4–7 cm × 3.5–6 cm, margin crenate, abaxially gray-green covered with pink powder. Male flowers fascicles 4, nearly sessile; female flowers fascicles 2–6. Follicles 2–4, pod-like. Fl. Apr, fr. Aug.

## 26. 悬铃木科

### 二球悬铃木 | *Platanus acerifolia*

**别名**：悬铃木、英桐、英国梧桐　　　　**科属**：悬铃木科悬铃木属

**形态特征**：落叶乔木，高可达35 m。树冠广卵圆形；树皮灰绿色，不规则地薄片状脱落，剥落后呈绿白色，光滑；幼枝密被褐色柔毛。单叶互生，近三角形，长10～24 cm，宽12～25 cm，上部掌状3～5裂，边缘疏生粗锯齿，中裂片长宽近相等，基部截形或心形，幼时叶两面被灰黄色星状绒毛，叶背毛尤密，后渐脱落。果序常2球一串，径约2.5 cm，宿存花柱刺毛状。花期4—5月，果期9—10月。

**习性、应用及产地分布**：温带强阳性树种。喜温暖、湿润气候，有一定耐寒性，在北方一些城市可露地栽培。对土壤的适应能力极强，耐干旱、瘠薄，亦耐水湿。萌芽力强，耐修剪；速生树种，根系浅，易风倒。树形优美，干皮光洁，冠大荫浓，对城市环境适应能力强，有"行道树之王"的美誉。为三球悬铃木与一球悬铃木的属内杂交种，现世界各地广泛种植。

**Description:** Trees, deciduous, up to 35 m tall. Bark gray green, cracking into irregular flakes, green white, smooth; young branchlets densely brown tomentose. Leaves alternate, triangular, 10−24 cm × 12−25 cm, principal veins 3 or 5, lobes entire or coarsely 1- or 2-dentate at margin, central lobe broadly triangular, abaxially densely pubescent, glabrescent with age. Infructescence capitate, ca. 2.5 cm in diam. Fl. Apr−May, fr. Sep−Oct.

## 27. 金缕梅科

### 枫香 | *Liquidambar formosana*

**别名**：枫树、路路通  **科属**：金缕梅科枫香属

**形态特征**：落叶乔木，高可达30 m。树皮灰色，浅纵裂，老时不规则深裂。叶互生，掌状3裂（萌芽枝的叶常为5～7裂），长6～12 cm，宽8～15 cm，先端渐尖，基部心形或截形，叶背被毛，后渐脱落，叶缘具细锯齿。花单性，雌雄同株，无花被；雄花序短穗状，多个排列成总状，雄蕊多数；头状雌花序生于雄花序下方叶腋，悬于细长花梗上。果序圆球形，木质，径3～4 cm，刺状萼片宿存，宿存花柱长1.5 cm。花期3—4月，果期10月。

**习性、应用及产地分布**：热带及亚热带阳性树种。喜温暖、湿润气候，耐干旱瘠薄，不耐水湿。萌蘖性强，主根粗长，抗风。树冠宽阔，树干通直，气势雄伟，可用于池畔、低山、丘陵地区营造风景林，大树移植困难，不耐修剪，一般不宜用作行道树。产于中国长江流域及其以南地区。中国南方著名的秋色叶树种。

**Description:** Trees, deciduous, up to 30 m tall. Bark gray. Leaves alternate, palmately 3-lobed, 6–12 cm × 8–15 cm, apex caudate-acuminate, cordate or truncate, densely pubescent, deciduous, margin glandular serrate. Male inflorescence a short spike, several arranged in a raceme, stamens many; female inflorescence 24–43 flowered. Infructescence globose, woody, 3–4 cm in diam., spiny sepals persistent, styles 1.5 cm. Fl. Mar–Apr, fr. Oct.

# 北美枫香 | *Liquidambar styraciflua*

**科属：** 金缕梅科枫香属

**形态特征：** 落叶乔木。树皮灰色，浅纵裂；小枝红褐色，通常有木栓质翅。叶互生，掌状5～7裂，叶长10～18 cm，叶柄长6.5～10 cm，叶背主脉有明显白簇毛，叶缘有锯齿；春、夏叶色暗绿，秋季叶色变为黄色或红色。果序圆球形，木质，刺状萼片宿存，宿存花柱长。

**习性、应用及产地分布：** 亚热带湿润气候树种。喜光照，适应性强，生长迅速，根深抗风，萌发能力强。树形优美，秋叶变为黄色、红色或紫色，宜作观赏树。原产于北美，有许多栽培品种，中国华东地区有引种。

**Description:** Trees, deciduous. Bark gray, branchlets red-brown. Leaves alternate, leaf blade palmately 5–7 lobed, 10–18 cm long, petioles 6.5–10 cm long, margin serrate. Leaves dark green in spring and summer, yellow, red, or purple in autumn. Infructescence globose, woody, spiny sepals persistent.

# 细柄蕈树 | *Altingia gracilipes*

**别名：** 细柄阿丁枫　　　　　　　　**科属：** 金缕梅科蕈树属

**形态特征：** 常绿乔木，高可达20 m。树皮灰褐色，片状剥落；小枝有柔毛。单叶互生，革质，卵状长椭圆形，长4～7 cm，先端有短尖，叶缘有尖锐细锯齿，叶面绿色，光亮，叶背无毛，网脉不明显；叶柄细长，2～3 cm。雄花序球状，多个成圆锥花序；雌花5～7朵成头状花序。果序倒圆锥形至球形，径1.5～2 cm，具5～6个木质蒴果。

**习性、应用及产地分布：** 喜光，喜温暖、湿润气候。不耐寒，侧根发达；萌芽力强，生长较快。常用于园林绿地观赏。分布于中国浙江南部、福建及广东。

**Description:** Trees, evergreen, up to 20 m tall. Bark gray brown, peeling off in strips; branchlets pubescent. Leaves alternate, leathery, leaf blade ovate-lanceolate, 4−7 cm long, apex caudate-acuminate, margin serrate, shiny, abaxially glabrous; reticulate veins indistinct on surfaces; petiole slender, 2−3 cm long. Male inflorescences globose, usually several arranged in a panicle; female inflorescences solitary or arranged in raceme, 5−7 flowered. Infructescences obconical, 1.5−2 cm in diam., with 5−6 woody capsules.

# 檵木 | *Loropetalum chinense*

**科属：** 金缕梅科檵木属

**形态特征：** 常绿灌木或小乔木，高4～9 m。小枝、幼叶、萼筒、子房和蒴果均被锈色星状毛。叶互生，革质，全缘，叶卵形或椭圆形，长2～5 cm，宽1.5～2.5 cm。花两性，3～8朵簇生于小枝顶端；花瓣先端圆钝，浅黄白色，长1～2 cm；花部4基数；萼筒倒锥形，与子房愈合，蒴果近卵形，长0.7～0.8 cm，褐色。花期3—4月，果期8—9月。

**习性、应用及产地分布：** 稍耐荫，喜温暖气候及酸性土壤，不耐寒。花繁密而显著，宜栽植于庭院观赏。分布于中国中部、南部及西南各省，亦见于日本及印度。

**Description:** Shrubs or small trees, 4–9 m tall. Branchlets, young leaves, calyces, and ovaries stellately rusty pubescent. Leaves alternate, leathery, entire, ovate or elliptic, 2–5 cm × 1.5–2.5 cm. Inflorescence a short raceme or nearly capitate, terminal, mostly on short lateral branches, 3–8 flowered; petals apex obtuse or rounded, pale yellow, 1–2 cm long; capsules ovoid or obovoid-globose, 0.7–0.8 cm long, brown. Fl. Mar–Apr, fr. Aug–Sep.

## 红花檵木 | *Loropetalum chinense* var. *rubrum*

**科属：** 金缕梅科檵木属

**形态特征：** 常绿灌木或小乔木。为檵木的变种，与原种的区别是叶片与花呈现暗紫红色。

**习性与应用：** 温带至亚热带树种。耐半荫，萌蘖力强，耐修剪。是南方优良的观花观叶树种，广泛用于绿地及色块构建。

**Description:** Shrubs or small trees. Variant of *Loropetalum chinense*, the difference from the original species is that the leaves and flowers are dark purple.

## 蜡瓣花 | *Corylopsis sinensis*

**科属：** 金缕梅科蜡瓣花属

**形态特征：** 落叶灌木，高2～5 m。小枝密被短柔毛。单叶互生，薄革质，倒卵状椭圆形，长5～9 cm，羽状脉，基部歪心形，叶背具星状毛，边缘具锐齿。花瓣5枚，柠檬黄色，先叶开放，芳香；花瓣如蜂蜡塑成，成下垂总状花序；蒴果卵球形，被褐色柔毛，种子黑色有光泽。花期3—4月，果期9—10月。

**习性、应用及产地分布：** 喜光，耐半荫，有一定耐寒能力。萌蘖力强，可天然更新。花期早而芳香，甚为秀丽。适于庭园观赏，或丛植于草地、林缘、路边或建筑物周边，颇具雅趣。产于中国长江流域以南各省。

**Description:** Shrubs, up to 2–5 m tall. Young branches and buds, pubescent or glabrous. Leaves alternate, leathery, leaf blade obovate to obovate-rounded, 5–9 cm long, base asymmetrical, cordate or subtruncate, abaxially stellately pubescent, margin serrate. Petals 5, lemon yellow, fragrant. Capsules obovoid-globose, brown pubescent, seed black, shiny. Fl. Mar–Apr, fr. Sep–Oct.

# 蚊母树 | *Distylium racemosum*

**科属：** 金缕梅科蚊母树属

**形态特征：** 常绿乔木，高可达25 m，栽培时常呈灌木状。树冠呈球形开展，幼枝具星状鳞毛。叶互生，厚革质，倒卵状长椭圆形至椭圆形，长3～7 cm，宽1.5～3.5 cm，先端钝或稍圆，基部宽楔形，全缘。总状花序长约2 cm，具星状毛，雌雄同序，雌花位于花序上部；雄蕊花药红色；花柱长6～7 mm。蒴果卵形，长1～1.3 cm，密生星状毛，开裂时顶端花柱宿存。花期4月，果期8—9月。

**习性、应用及产地分布：** 亚热带树种。喜光，也耐荫；对土壤要求不严。萌芽、发枝力强，耐修剪。对烟尘及多种有毒气体抗性很强。叶常生虫瘿，影响观赏。树形整齐，枝叶密集，宜植于路旁、庭前草坪。对多种有毒气体抗性强，防尘及隔音效果好，可栽作绿篱和防护林带，丛植、片植作为分隔空间效果亦佳。产于中国广东、福建、台湾、浙江等地，长江流域城市园林中常见栽培。

**Description:** Trees, evergreen, up to 25 m tall. Crown spreading globose, young branches stellately lepidote. Leaves alternate, leathery, blade elliptic or obovate-elliptic, 3−7 cm × 1.5−3.5 cm, apex obtuse or subacute, base broadly cuneate, margin entire. Inflorescences 2 cm, stellately pubescent, monoecious; anthers red, stylets 6−7 mm. Capsules ovoid, 1−1.3 cm long, densely stellately pubescent. Fl. Apr, fr. Aug−Sep.

# 小叶蚊母树 | *Distylium buxifolium*

**科属**：金缕梅科蚊母树属

**形态特征**：常绿灌木，高 1～2 m。枝灰褐色。单叶互生，薄革质，倒披针形至长圆状倒披针形，长 3～5 cm；叶面绿色，侧脉不明显，在叶背略突起，全缘；叶柄极短。穗状花序腋生，长 1～3 cm，苞片线状披针形。蒴果卵圆形，长 7～8 mm，宿存花柱长 1～2 mm。

**习性、应用及产地分布**：喜光，稍耐荫，喜温暖湿润气候，喜酸性、中性土壤，萌芽发枝力强，耐修剪。优良的色块、矮篱植物，优良的水土保持植物，可用于滨水绿地、岸坡栽植。分布于中国四川、湖北、湖南、福建、广东及广西等省区。常生于山溪旁或河边。

**Description:** Shrubs, evergreen, 1～2 m tall. Branches gray brown. Leaves alternate, thin leathery, leaf blade oblanceolate to oblong-lanceolate or obovate, 3–5 cm long, green, reticulate veins obscure on both surfaces, entire. Petiole short, 1–3 cm long. Capsules ovate-rounded, 7–8 mm long, persistent stylet 1–2 mm long.

# 28. 虎耳草科

## 山梅花 | *Philadelphus incanus*

**科属：** 虎耳草科山梅花属

**形态特征：** 落叶灌木，高3～5 m。树皮褐色，薄片状剥落，枝具白色髓心；1年生小枝被柔毛，后脱落。单叶对生，基部三出脉或五出脉，叶卵形至卵状长椭圆形，长4～8 cm，叶面疏生直立刺毛，叶背密生短毛，缘具疏齿。总状花序具花5～11朵，白色，芳香，形似梅花，径2～3 cm；萼片、花瓣均为4枚；花梗及花萼均被毛。蒴果倒卵形，径5～7 mm。花期6—7月，果期8—9月。

**习性、应用及产地分布：** 温带树种。性强健，稍耐荫，较耐寒，耐旱，不耐水湿。生长速度快，萌蘖力强。枝叶稠密，花白清香，花期长。宜栽植于庭园及公园，也可与建筑物、山石等搭配点缀，丛植或片植于林缘、山坡及草地。产于中国河南、山西、江苏、陕西、甘肃南部、湖北、湖南、江西、四川等地。

**Description:** Shrubs, deciduous, 3–5 m tall. Bark brown, peeling off in flakes; 1st-year branches pubescent, decisuous when age. Leaves opposite, base veins 3–5, leaf blade ovate to obovate-elliptic, 4–8 cm long, adaxially bristly, abaxially densely pubescent, margin sparsely serrate. Racemes 5–11 flowered, petals white, 2–3 cm in diam; sepals and petals 4, pedicels and sepals are pubescent. Capsules obovate, 5–7 mm in diam. Fl. Jun–Jul, fr. Aug–Sep.

# 溲疏 | *Deutzia scabra*

**科属：** 虎耳草科溲疏属

**形态特征：** 落叶灌木，高2.5 m。树皮薄片状剥落；小枝红褐色，中空，幼时被星状柔毛。单叶对生，叶卵状椭圆形，长4～6 cm，宽2～4 cm，两面被星状毛，粗糙，边缘具不明显的细锯齿。直立圆锥花序生于小枝顶端，花两性，花瓣5枚；花梗密被星状毛，花萼裂片较萼筒短；花瓣白色或外面略带粉红色，稍有香气，外被星状毛。蒴果近球形，径3～5 mm。花期5—6月，果期7—8月。

**习性、应用及产地分布：** 暖温带至亚热带阳性树种。性强健，喜温暖气候，较耐寒，耐旱。萌芽力强，耐修剪。夏季白花繁密而素雅，花期长，其重瓣、斑叶变种更具观赏性，庭园习见栽培。宜丛植于草坪、建筑物及路旁、林缘、山坡，也可作花篱及岩石园种植材料。原产于日本，中国华北、华东地区常见栽培观赏。

**Description:** Shrubs, deciduous, 2.5 m tall. Bark peeling off in flakes; branchlets red-brown, hollow, stellate hairy when young. Leaves opposite, ovate-elliptic, 4–6 cm × 2–4 cm, stellate hairy on both surfaces, rough, margin unobvious serrate. Erect panicle, petals 5; pedicels stellate pubescent; petals white or pink, fragrant, stellate pubescent. Capsules subglobose, 3–5 mm in diam. Fl. May–Jun, fr. Jul–Aug.

# 雪球冰生溲疏 | *Deutzia gracilis* 'Nikko'

**科属：** 虎耳草科溲疏属

**形态特征：** 落叶小灌木，株高0.4～0.6 m。冠幅1.0～1.2 m，枝条多且柔软。单叶对生，披针形，长3～6 cm，边缘有锯齿，亮绿色，秋季变为红色。圆锥花序长10～15 cm，小花白色，重瓣，花径约1.5 cm，极为繁茂。花期4～5月。

**习性、应用及产地分布：** 此品种因其株型低矮丰满，开花繁茂而闻名，是园林绿化中的优良观赏树种，可散植、群植观赏。产于中国湖北，华北、华东各省有栽培，朝鲜半岛也有分布。

**Description:** Shrubs, deciduous, 0.4–0.6 m tall. Crown 1.0–1.2 m, branches many and soft. Leaves opposite, lanceolate, 3–6 cm long, margin serrate，bright green, turn red in autumn. Panicle 10–15 cm, petals white, double, ca. 1.5 cm in diam., lush. Fl. Apr–May.

## 绣球花 | *Hydrangea macrophylla*

**别名：** 大八仙花、紫绣球、粉团花、八仙绣球　　　**科属：** 虎耳草科绣球属

**形态特征：** 落叶灌木，高3～4 m。小枝粗壮，长形皮孔明显。叶对生，叶大而有光泽，倒卵形至椭圆形，长7～20 cm，先端渐尖或急尖，基部近圆形，两面无毛或仅下面脉上被毛，缘有粗锯齿。顶生伞房花序近球形，径12～20 cm；花序中几乎全部是大型不育花，蓝色或粉红色。蒴果长陀螺状，具宿存花柱。花期6—7月。

**习性、应用及产地分布：** 亚热带阴性树种。喜温暖、湿润气候，耐寒性不强；花色会因土壤酸碱度的变化而改变，酸性土时花白色至蓝色，碱性土时花则呈粉红色，是土壤酸碱性的指示植物。萌蘖力强，病虫害较少。植株健壮，花序大而美丽，中式、日式和西式园林均适宜栽植。产于中国长江流域以南各省区，朝鲜、日本亦有分布。

**Description:** Shrubs, deciduous, 3–4 m tall. Branchlets thick, long lenticels obvious. Leaves opposite, large and shiny, obovate to elliptic, 7–20 cm long, base subrounded, glabrous on both surfaces, margin rough serrate. Corymb subglobose, 12–20 cm in diam., blue or pink. Capsules long turbinate, stylet persistent. Fl. Jun–Jul.

# 银边八仙花 | *Hydrangea macrophylla* var. *normalis* 'Maculata'

**科属：** 虎耳草科绣球属

**形态特征：** 落叶灌木，此种为绣球花的变种。与原种不同之处为花序近扁平型，大部分为两性可育花，只在花序边缘有少数大型不育花；不育花仅有扩大的萼片4枚，呈花瓣状，卵形；花粉红色、蓝色或白色。产于日本及中国浙江。

**Description:** Shrubs, deciduous, variant of *Hydrangea macrophylla*. Inflorescence subflat, most bisexual fertile flowers, only a few large sterile flowers on inflorescence margin; sterile flowers having only 4 enlarged sepals, petalous, ovate; pollen red, blue or white. Original from Japan and Zhejiang Province of China.

# 29. 海桐花科

## 海桐 | *Pittosporum tobira*

**别名：** 海桐花、七里香　　　　　　**科属：** 海桐花科海桐花属

**形态特征：** 常绿灌木，高 2～6 m。树冠圆球形，幼枝被褐色柔毛。叶厚革质，倒卵状椭圆形，长 4～9 cm，宽 1.5～4 cm，先端圆钝或微凹，基部窄楔形，上面有光泽，全缘，边缘反卷。伞形或伞房状花序顶生；花白色，后变淡黄色，花瓣离生，芳香，径约 1 cm；苞片、小苞片与萼片均被毛。蒴果球形或倒卵状球形，径 1～1.2 cm，有棱角，多少被毛，果皮木质。花期 5—6 月，果期 9—10 月。

**习性、应用及产地分布：** 亚热带阳性树种。喜温暖、湿润气候，不耐水湿。对土壤要求不严。萌芽力强，耐修剪；生长较快。树冠球形，叶色浓绿而有光泽，经冬不凋；初夏花朵美丽、芳香；入秋果熟开裂时露出红色种皮，亦很美观，是南方习见绿化树种。庭园中常植作绿篱或修剪成球、孤植、丛植于草坪边缘、林缘、树坛、花坛，或对植于门旁。产于中国东南沿海等地，长江流域以南庭园习见栽培；朝鲜、日本亦有分布。

**Description:** Shrubs, evergreen, 2–6 m tall. Crown globose-rounded, branchlets brown pubescent. Leaves leathery, obovate or obovate-lanceolate, 4–9 cm × 1.5–4 cm, apex rounded or obtuse, base narrowly cuneate, shiny adaxially, margin entire, revolute. Inflorescences umbellate or corymbose; petals, white, becoming yellow later; bracts, bractlets, sepal pubescent. Capsule globose, 1–1.2 cm in diam., angular, pubescent, pericarp woody. Fl. May–Jun, fr. Sep–Oct.

## 斑叶海桐 | *Pittosporum tobira* 'Variegata'

**科属：** 海桐花科海桐花属

**形态特征：** 常绿灌木，高 2～6 m。形态与原种相似，叶面具有不规则白色斑纹。

**Description:** Shrubs, evergreen, 2–6 m tall. Similar to the original species, leaves present irregular white stripes.

# 30. 蔷薇科

## 麻叶绣线菊 | *Spiraea cantoniensis*

**科属：** 蔷薇科绣线菊属

**形态特征：** 落叶灌木，高1.5 m。小枝细长，暗红色，呈拱形弯曲。叶菱状披针形至菱状长圆形，长3～5 cm，宽1.5～2 cm，叶面暗绿色，叶背青蓝色，两面光滑，边缘自中部以上具缺刻状锯齿；羽状脉。花10～30朵集成半球状伞形花序，生于去年生小枝顶端；花瓣白色；雄蕊20～28枚，稍短于花瓣或与花瓣近等长。蓇葖果直立，花柱顶生，具宿萼。花期4—5月，果期7—9月。

**习性、应用及产地分布：** 庭园常见的观花灌木。产于中国华南、华东地区，日本亦有分布。

**Description:** Shrubs, deciduous, 1.5 m tall. Branchlets slender, dark red, arching. Leaves alternate, rhombic-lanceolate to rhombic-oblong, 3–5 cm × 1.5–2 cm, dark green adaxially, leaf blade gray-blue abaxially, margin incised serrate above middle of leaf, apex acute, glabrous, pinnately veined. Cyme inflorescences, 10 to 30 flowers froming hemispherical; petals white, stamens 20–28, slightly shorter than to nearly equaling petals. Follicles straightly spreading, styles terminal. Fl. Apr–May, fr. Jul–Sep.

## 粉花绣线菊 | *Spiraea japonica*

**别名**：日本绣线菊　　　　　　　**科属**：蔷薇科绣线菊属

**形态特征**：落叶灌木，高1.5 m。枝条开展、细长，小枝光滑。叶卵状披针形，长2～8 cm，宽1～3 cm，先端急尖至短渐尖，基部楔形，上面沿脉微被短柔毛，下面略带白粉，边缘具重锯齿或单锯齿；叶柄具短柔毛。复伞房花序平截，生于1年生枝端；花瓣粉红色、紫红色，径4～7 mm。蓇葖果卵状椭圆形，半张开，花柱顶生，萼片宿存。花期6—7月，果期8—9月。

**习性、应用及产地分布**：抗逆性强（抗寒、耐盐碱、抗病虫害、抗污染），耐修剪。产于日本、朝鲜，中国吉林、辽宁、河北、河南、山东各地均有栽培。

**Description:** Shrubs, deciduous, 1.5 m tall. Branchlets slender, smooth. Leaves alternate, rhombic-lanceolate to rhombic-oblong, 2–8 cm × 1–3 cm, margin incised serrate above middle of leaf, apex acute, glabrous, pinnately veined. Cyme inflorescences, hemispherical; petals pink to purple-red, 4–7 mm in diam., nearly equal in size. Follicles straightly spreading, styles terminal. Fl. Jun-Jul, fr. Aug-Sep.

# 珍珠花 | *Spiraea thunbergii*

**别名：** 喷雪花、雪柳、珍珠绣线菊　　　　**科属：** 蔷薇科绣线菊属

**形态特征：** 落叶灌木，高 1.5 m。枝纤细而密生，开展并拱曲，有棱角，幼时被短柔毛，老时无毛。单叶互生，线状披针形，长 2.5～4 cm，中部以上有尖锯齿，两面无毛，羽状脉；叶柄长 1～2 mm，有柔毛。伞形花序无花序梗，花小，白色，径 6～8 mm。蓇葖果具宿存花柱与萼片。花期 4—5 月，果期 6—7 月。

**习性、应用及产地分布：** 花朵密集如积雪，叶片薄细如鸟羽，叶片秋季转变为橘红色，甚为美丽。原产于中国华东地区，日本也有分布。

**Description:** Shrubs, deciduous, 1.5 m tall. Branches spreading, arching, slender, angled, pubescent at first, finally glabrous. Leaves alternate, leaf blade linear-lanceolate, 2.5–4 cm long, margin sharply serrate above middle, glabrous on both surfaces, pinnately veined; petiole 1–2 mm, pubescent. Umbels sessile, petals white, 6–8 mm in diam. Fl. Apr–May, fr. Jun–Jul.

# 风箱果 | *Physocarpus amurensis*

**科属：** 蔷薇科风箱果属

**形态特征：** 落叶灌木，高可达3 m。树皮纵向剥裂，枝条开展；小枝稍弯曲，幼时紫红色。叶三角状卵形，3～5浅裂，长3.5～5.5 cm，宽3～5 cm，先端急尖或渐尖，基部近心形，叶背微被星状毛与短柔毛，边缘具重锯齿。伞形总状花序，径3～4 cm，花梗长1～1.8 cm；花白色，径约1 cm，花药紫色。蓇葖果成熟时膨大，径约1 cm，外微被星状柔毛；种子2～5粒，黄色，有光泽。花期6月，果期7—8月。

**习性、应用及产地分布：** 温带阳性树种。适应性强，耐寒性强，耐干旱瘠薄，但不耐水渍。萌芽力强，耐修剪。夏季白花密集，初秋红果累累，颇为美观，为夏秋观花、赏果灌木。可孤植、丛植于坡地、丛林边缘、路旁、亭台及假山周围，也可作花篱。中国东北地区多有栽培。朝鲜北部及俄罗斯远东地区亦有分布。

**Description:** Shrubs, deciduous, 3 m tall. Bark vertical fissures, branches spreading; branchlets arching, purple-red. Leaf blade triangular-ovate to broadly ovate, 3.5–5.5 cm × 3–5 cm, apex acute or acuminate, base cordate or subcordate, abaxially stellate hairy and pubescent, margin doubly serrate. Umbel inflorescence, 3–4 cm in diam., pedicel 1–1.8 cm; petals white, ca. 1 cm in diam., anthers purple. Follicles swell when mature, ca. 1 cm in diam., stellate pubescent. Seeds 2–5, yellow, shiny. Fl. Jun, fr. Jul–Aug.

# 平枝栒子 | *Cotoneaster horizontalis*

**别名**：铺地蜈蚣　　　　　　　　　　**科属**：蔷薇科栒子属

**形态特征**：落叶或半常绿匍匐灌木，高约0.5 m。枝水平开展呈整齐2列状，幼枝被黄褐色粗毛，老时脱落。叶厚革质，近圆形或宽椭圆形，长5～15 mm，宽4～9 mm，先端急尖，基部楔形，背面疏生平贴细毛，全缘。花1～2簇生，近无柄；萼筒钟状，外面有稀疏柔毛；花瓣粉红色或白色，直立；雄蕊短于花瓣；花柱离生。果近球形，径4～6 mm，熟时鲜红色，经冬不落。花期5—6月，果期9—10月。

**习性、应用及产地分布**：亚热带阳性树种。适生于温暖、湿润气候，较耐寒，耐干旱瘠薄，不耐涝。枝繁叶密，深秋叶色红亮，果实累累，为优良的观赏地被植物，宜丛植于建筑前及园路两旁、假山、叠石及溪畔，或点缀于小型庭院、公园。产于中国西北、中南及西南地区。

**Description:** Shrubs, deciduous or semi-evergreen, 0.5 m tall. Branchlets yellowish brown, initially strigose. Leaves leathery, suborbicular or broadly elliptic, rarely obovate, 5–15 mm × 4–9 mm, base cuneate, apex acute, abaxially sparsely accumbent pubescent, entire. Inflorescences 1- or 2-flowered, pedicel short to nearly absent; hypanthium campanulate, abaxially sparsely pubescent; petals erect, pink, reddish, or whitish; stamens shorter than petals. Fruit bright red, subglobose or ellipsoid, 4–6 mm in diam. Fl. May–Jun, fr. Sep–Oct.

## 火棘 | *Pyracantha fortuneana*

**别名：** 火把果、救军粮　　　　**科属：** 蔷薇科火棘属

**形态特征：** 常绿灌木，高约3 m。枝具棘刺，幼时被锈色短柔毛。叶倒卵形至倒卵状长椭圆形，长1.5～6 cm，宽0.5～2 cm，先端圆钝或微凹，基部楔形，上面有光泽，边缘具圆钝锯齿，近基部全缘。复伞房花序，径3～4 cm；萼筒钟状；花瓣白色，近圆形；雄蕊20枚。梨果近球形，径约5 mm，橘红色或深红色。花期4—5月，果期9—10月。

**习性、应用及产地分布：** 喜光，极耐干旱瘠薄，不耐寒。萌芽力强，耐修剪。初夏白花繁密，入秋果红似火，经久不凋，颇为美观，是传统的观果树种。在园林绿地中丛植、篱植、孤植皆宜。产于中国黄河流域以南及西南地区。

**Description:** Shrubs, evergreen, 3 m tall. Lateral branches short, spiny, young branchlets rusty pubescent. Leaf blade obovate or obovate-oblong, 1.5–6 cm × 0.5–2 cm, apex obtuse or emarginate, base cuneate, subbase entire. Complex cymes inflorescences, 3–4 cm in diam.; hypanthium campanulate, petals white, subrounded. Pome orangish red or dark red, subglobose. Fl. Apr–May, fr. Sep–Oct.

# 小丑火棘 | *Pyracantha fortuneana* 'Harlequin'

**科属：** 蔷薇科火棘属

**形态特征：** 常绿灌木，此种是火棘的栽培变种。叶片具斑纹，似小丑花脸，故名小丑火棘。冬季叶片粉红色，观赏价值更高。日本选育而成。

**Description:** Shrubs, evergreen, variant of *Pyracantha fortuneana*. Leaves with irregular patches of color. Winter leaves pink, possessing high ornamental value. Selected in Japan.

## 山楂 | *Crataegus pinnatifida*

**别名**：酸楂、五月花　　　　　　　**科属**：蔷薇科山楂属

**形态特征**：落叶小乔木，高可达6 m。常具枝刺。叶广卵形或三角状卵形，两侧3～6对羽状深裂，长5～12 cm，宽4～7.5 cm，先端短渐尖，基部平截或宽楔形，边缘具不规则尖锐重锯齿；叶柄细，长2～6 cm；侧脉直伸至裂片先端；托叶镰形，具齿。伞房花序顶生，花梗被长柔毛；苞片膜质；雄蕊20枚，短于花瓣。梨果近球形，径约1.5 cm，深红色，具白色皮孔。花期5—6月，果期9—11月。

**习性、应用及产地分布**：温带树种。树势强健，适应能力强。喜光、耐寒、耐旱，对土壤要求不严，根系发达，萌芽性强。树冠整齐，枝叶繁茂，花艳果红，是观花、观果树种，可作庭荫树和园路树。分布在中国东北、华北、江苏及朝鲜、俄罗斯西伯利亚地区。

**Description:** Trees, deciduous, 6 m tall. Leaves alternate, broadly ovate or triangular-ovate, with 3–6 pairs of lobes, 5–12 cm × 4–7.5 cm, apex shortly and acuminate, base truncate or broadly cuneate, margin sharply irregularly doubly serrate; petiole 2–6 cm; lateral veins usually extending to apices of lobes, margin serrate. Inflorescences cymose terminal, bracts membranous. Pome subglobose, nearly 1.5 cm in diam., dark red, with white lenticels. Fl. May–Jun, fr. Sep–Nov.

# 枇杷 | *Eriobotrya japonica*

**科属：**蔷薇科枇杷属

**形态特征：**常绿小乔木，高10 m。小枝、叶下面、叶柄和花序均密生灰棕色或锈色绒毛。大型单叶互生，厚革质，披针形、长倒卵形或长椭圆形，长12～30 cm，宽3～9 cm，先端急尖或渐尖，基部楔形或渐狭成叶柄，上面多皱而有光泽，仅上部边缘具疏锯齿；羽状脉明显，11～21对，侧脉直达齿尖。圆锥花序长10～19 cm；花白色，径1.2～2 cm，芳香；萼筒浅杯状；花瓣内面被绒毛。梨果近球形或长圆形，径2～2.5 cm，黄色或橘黄色，外被锈色柔毛，后脱落。花期10—12月，果期翌年5—6月。

**习性、应用及产地分布：**喜光，适生于温暖、湿润气候，稍耐寒。叶形酷似中国传统乐器琵琶而得名"枇杷"。叶大荫浓，四季碧叶，多植为庭荫树。中国长江流域以南有栽培，日本、东南亚地区亦有分布。

**Description:** Small trees, evergreen, 10 m tall. Branchlets, leaf back, petiole and peduncle densely rusty or grayish rusty tomentose. Leaves leathery, blade lanceolate, obovate, or elliptic-oblong, 12–30 cm × 3–9 cm, apex acute or acuminate, base cuneate adaxially lustrous, rugose, margin entire basally, upper margin sparsely serrate only. Lateral veins 11 or 21 pairs. Panicle 10–19 cm; flowers white, fragrant, 1.2–2 cm in diam.; hypanthium shallowly cupular; petals inside tomentose. Pome subglobose, 2–2.5 cm in diam., yellow. Fl. Oct–Dec, fr. May–Jun.

# 石楠 | *Photinia serrulata*

**科属：** 蔷薇科石楠属

**形态特征：** 常绿小乔木，高6～12 m。小枝灰褐色，无毛。叶革质，长椭圆至倒卵状椭圆形，长9～22 cm，宽3～6.5 cm，先端渐尖，基部圆形或宽楔形，幼叶红色，有光泽，边缘具细腺齿，近基部全缘；中脉显著。复伞房花序，径10～16 cm；花白色，径6～8 mm，花瓣近圆形。雄蕊10枚；子房顶部被绒毛，花柱2～3个，基部连合。果球形，径5～6 mm，红色。花期4—5月，果期10月。

**习性、应用及产地分布：** 亚热带阳性树种。喜温暖、湿润气候，较耐寒，不耐水湿。萌芽力强，耐修剪；对有毒气体抗性强。枝叶浓密，早春嫩叶鲜红，春尽白花成簇，秋冬累累赤实，季相变化丰富，可广泛用于庭园、公园及园路旁，尤宜配植于整形式园林中。产于中国秦岭以南至华南、西南地区，日本、菲律宾、印度尼西亚亦有分布。

**Description:** Trees, evergreen, 6–12 m tall. Branchlets brownish gray, glabrous. Leaf leathery, blade elliptic or obovate-elliptic, 9–22 cm × 3–6.5 cm, apex acuminate, base rounded or broadly cuneate, red when young, margin sparsely or inconspicuously toothed or entire; midvein obvious. Compound corymbs, 10–16 cm in diam.; petals white, 6–8 mm in diam., suborbicular. Stamens 10; ovary terminal tomentose, styles 2–3, connate at base. Fruit globose, 5–6 mm in diam., red. Fl. Apr–May, fr. Oct.

# 红叶石楠 | *Photinia* × *fraseri*

**科属：**蔷薇科石楠属

**形态特征：**是石楠与光叶石楠（*Photinia glabra*）育成的杂交种常绿小乔木，高4～6 m。茎直立；叶片革质，长椭圆形至倒卵状披针形，下部叶绿色或带紫色，上部嫩叶鲜红色或紫红色；新梢和嫩叶春、秋、冬三季呈红色，色彩亮丽持久，低温色更佳。花期4—5月，果期10月。

**习性、应用及产地分布：**性强健，较耐寒，对土壤要求不严，耐修剪，可用作绿篱植物，观赏性及适应性明显优于同色系红花檵木和'紫叶'小檗。欧美和日本广泛应用，被誉为"红叶绿篱之王"。

**Description:** Small trees, evergreen, 4–6 m tall. Stem erect; leaf leathery, blade elliptic or obovate-lanceolate, green or purplish lower, bright red or red purplish upper; new branchlets and leaves red in spring, autumn and winter. Fl. Apr–May, fr. Oct.

# 椤木石楠 | *Photinia davidsoniae*

**科属：** 蔷薇科石楠属

**形态特征：** 常绿乔木，高6～15 m。树干、枝条常有刺，幼枝发红。叶革质，长圆形或倒卵状披针形，长5～15 cm，先端急尖或渐尖，基部楔形，边缘具细腺锯齿，稍反卷；叶柄长0.8～1.5 cm。花多而密，成顶生复伞房花序，花序梗及花梗贴生短柔毛；花径1.0～1.2 cm，花瓣近圆形。梨果球形或卵形，径7～10 mm，橙红色。花期5月，果期9—10月。

**习性、应用及产地分布：** 喜光、耐寒、耐旱、不耐水湿。树冠整齐，可用于刺篱与造型，适用于园林绿化与工矿区配植。产于中国长江以南至华南地区海拔600～800 m灌丛中，越南、缅甸、泰国亦有分布。

**Description:** Trees, evergreen, 6–15 m tall. Trunk and branches with thorny, branchlets red when young. Leaves leathery, blade elliptic or obovate-elliptic, 5–15 cm, apex acuminate, base cuneate, margin serrate; petiole 0.8–1.5 cm. Compound corymbs terminal, peduncles pedicels short-villous; flowers 1.0–1.2 cm in diam., petals suborbicular. Pome globose or ovate, 7–10 mm in diam., orange red. Fl. May, fr. Sep–Oct.

# 日本贴梗海棠 | *Chaenomeles japonica*

**别名**：倭海棠　　　　　　　　　**科属**：蔷薇科木瓜属

**形态特征**：落叶灌木，通常不及1 m。枝紫褐色，开展，具枝刺。叶长卵形至椭圆形，长3～6 cm，基部楔形或宽楔形，上面有光泽，边缘具圆钝锯齿；托叶肾形或半圆形。花3～5朵簇生于2年生枝上，径3～5 cm；花梗粗短；花瓣火焰色或亮橘红色。梨果卵形或近球形，径4～6 cm，黄色。花期3—5月，果期9—10月。

**习性、应用及产地分布**：温带树种，喜光、耐寒、耐旱，不耐积水。原产于日本，中国各地庭园见栽培观赏。

**Description**: Shrubs, deciduous, less than 1 m. Branches purple-brown with thorns. Leaves long ovate to elliptic, 3–6 cm long, base cuneate or broadly cuneate, glaucously. Flowers 3–5-fascicled on second-year branchlets, 3–5 cm in diam.; peduncle, short and thick; petals, flame-colored or bright orange-red. Pome nearly spherical in shape with a diam. of 4–6 cm, yellow when mature. Fl. Mar–May, fr. Sep–Oct.

# 木瓜 | *Chaenomeles sinensis*

**科属：** 蔷薇科木瓜属

**形态特征：** 落叶小乔木，高可达10 m。树皮斑驳，薄片状剥落，枝无刺。单叶互生，革质，卵状椭圆形，长5～8 cm，缘有芒状锐齿；幼叶密被柔毛，后脱落。花单生，粉红色，径3～4 cm；梨果椭球形，长10～15 cm，深黄色，有香气。花期4—5月。

**习性、应用及产地分布：** 喜光，喜温暖、湿润气候，不耐寒。本种花红果香，干皮斑驳秀丽，常植于庭园观赏。产于中国东部及中南地区。

**Description:** Small trees, deciduous, 10 m tall. Leaves alternate, leathery, elliptic-ovate, 5-8 cm, margin spiniform teeth. Flowers solitary, pink, 3-4 cm in diam. Pome ellipsoid, 10-15 cm, dark yellow, fragrant. Fl. Apr-May.

# 垂丝海棠 | *Malus halliana*

**别名：** 垂枝海棠　　　　　　　　　　**科属：** 蔷薇科苹果属

**形态特征：** 落叶小乔木，高可达8 m。树皮灰褐色、光滑；枝条开展；幼枝褐色，疏生短柔毛。叶椭圆形至长椭圆形，长3.5～8 cm，宽2.5～4.5 cm，先端长渐尖，基部楔形至近圆形，上面有光泽，下面被短柔毛，边缘有圆钝细锯齿；叶柄细长。花4～6组成伞形总状花序，花梗细弱，长2～4 cm，下垂；萼片三角状卵形，先端钝；花多半重瓣，未开时红色，开后渐变为粉红色；雄蕊20～25枚。梨果近球状，径6～8 mm，黄绿色略带紫色，萼片脱落。花期3—4月，果期9—10月。

**习性、应用及产地分布：** 亚热带阳性树种。喜光，不耐荫，也不耐寒，喜温暖、湿润气候，对土壤要求不严。花繁色艳，花朵下垂，甚为美丽，是著名的庭园观赏花木。产于中国西南部，长江流域至西南各地均有栽培。

**Description:** Small trees, deciduous, 8 m tall. Bark taupe, smooth; branchlets brown, puberulous when young. Leaves elliptic, 3.5–8 cm × 2.5–4.5 cm, lustrous puberulous, apex long acuminate, base cuneate or subrounded, margin obtusely serrulate, petiole long and slender. Corymb 4–6 cm in diam., 4–6 flowered; pedicel slender, 2–4 cm, pendulous; sepals triangular-ovate, apex obtuse; petals often red and gradually become pink. Pome globose, 6–8 mm in diam., yellow-green and purplish, pyriform or obovoid, sepals caducous. Fl. Mar–Apr, fr. Sep–Oct.

# 玫瑰 | *Rosa rugosa*

**科属：** 蔷薇科蔷薇属

**形态特征：** 落叶丛生灌木，高2 m。枝灰褐色，密生刚毛与皮刺。复叶具小叶5～9枚，厚革质，椭圆形或椭圆状倒卵形，长2～5 cm，宽1.5～2.5 cm，先端急尖或圆钝，基部圆形或宽楔形，边缘具钝锯齿，上面深绿色，叶脉下陷而多皱，叶背灰绿色，密被绒毛；托叶大部附着于叶柄，边缘具腺齿。花单生或数朵聚生于叶腋，径6～8 cm，花梗密被绒毛和腺毛；花瓣紫红色，重瓣至半重瓣，浓香。果扁球形，径2～2.5 cm，熟时红色，萼片宿存。花期5—6月，果期9—10月。

**习性、应用及产地分布：** 温带阳性树种。耐寒，耐旱，但不耐水渍。萌蘖力强。最宜作花篱、花境、花坛及玫瑰专类园栽植，也是现代月季之亲本。产于中国、日本和朝鲜，中国辽宁、山东等地有分布，现世界各地广泛栽培。

**Description:** Shrubs deciduous, 2 m tall. Branchlets taupe, tomentose and prickles dense. Leaves pinnate; leaflets 5–9, leathery, elliptic or elliptic-obovate, 2–5 cm × 1.5–2.5 cm, apex acute or rounded-obtuse, base rounded or broadly cuneate, margin acutely serrate, adaxially dark-green, rugose due to concave veins, abaxially gray-green, tomentose; stipules mostly adnate to petiole, margin glandular-pubescent. Flower solitary, or several and fasciculate, axillary, 6–8 cm in diam.; petals purple-red, fragrant. Fruit, 2–2.5 cm in diam., red. Fl. May–Jun, fr. Sep–Oct.

# 月季花 | *Rosa chinensis*

**别名：**月月花、月月红　　　　　　　　**科属：**蔷薇科蔷薇属

**形态特征：**常绿或半常绿灌木，高1～2 m。茎常具钩状皮刺。复叶具小叶3～5枚，小叶宽卵形至卵状椭圆形，长2.5～6 cm，宽1～3 cm，先端尖，基部圆形或宽楔形，两面近无毛，叶面鲜绿色而有光泽，边缘具尖锐细锯齿；托叶大部贴生于叶柄，仅顶端分裂成耳状。花单生或数朵簇生，径约4～6 cm，花梗常被腺毛；萼筒近球形，萼片三角状卵形，内面和边缘被短绒毛；花深红色、粉红色、稀白色，重瓣，微香；花柱离生。果卵球形或梨形，长1.5～2 cm，黄红色，萼片宿存。花期4—9月，果期6—11月。

**习性、应用及产地分布：**温带及亚热带阳性树种。适生于温暖、湿润气候，耐干旱，不耐水涝。花色艳丽，花期长，适应性强，丛植、群植于庭院、园路角隅、假山、草坪、花境等处，或作专类园。产于中国，现世界各地普遍栽培观赏。

**Description:** Shrubs, evergreen, 1–2 m tall. Prickles abundant. Leaves pinnate; leaflets 3–5, broadly ovate or ovate-oblong, 2.5–6 cm × 1–3 cm, apex acuminate, base subrounded or broadly cuneate, both surfaces subglabrous, margin acutely serrate, stipules mostly adnate to petiole, only top part auriculate. Flowers solitary or fasciculate, 4–6 cm in diam., pedicel glandular-pubescent; sepals triangular-ovate, adaxially and margin pubescent; flower dark-red, pink, rarely white. Fruit ovoid or pyriform, yellow-red. Fl. Apr–Sep, fr. Jun–Nov.

## 安吉拉月季 | *Rosa hybrida* 'Angela'

**科属：** 蔷薇科蔷薇属

**形态特征：** 安吉拉是丰花月季品种之一，高0.8～1.5 m。枝条粗壮，少刺，半直立生长。花常集生于枝端，花冠杯状，粉红色，径约4 cm，无香。花瓣10～35枚。多季节连续盛开。

**习性与应用：** 耐寒，有活力。可用于盆栽、花坛和边界，切花或花园。

**Description:** Variant of floribunda roses, 0.8–1.5 m tall. Branchlet strong, rare spinescent, semi-erect. Flowers clustered at branch terminal, cupular, pink, ca. 4 cm in diam., not fragrant, petals 10–35, continuous bloom in multi-season.

# 粉团蔷薇 | *Rosa multiflora* var. *cathayensis*

**别名：** 粉花蔷薇  **科属：** 蔷薇科蔷薇属

**形态特征：** 落叶灌木，高可达3 m。皮刺常生于托叶下。羽状复叶互生，小叶5～9枚，倒卵状椭圆形，缘具尖锯齿，叶背有柔毛；托叶篦齿状附着于叶柄基部。伞房花序，花单瓣，径3～4 cm，粉红至玫红色，花梗常具腺毛。果红色。花期5—6月。

**习性、应用及产地分布：** 性强健，喜光、耐寒、耐旱。可栽作花篱，或植于水边、林缘观赏。产于中国长江流域各地，南至两广、云南、四川、贵州等地。

**Description:** Shrubs, deciduous, 3 m tall. Prickles abundant. Pinnate leaves alternate, leaflets 5–9, obovate-elliptic, margin acutely serrate, abaxially tomentose; stipules comb shaped and attached to the base of the petiole. Corymbs, single petal, 3–4 cm in diam., pink and rose-red, pedicel glandular-pubescent. Fruit red. Fl. May–Jun.

# 七姊妹 | *Rosa multiflora* var. *carnea*

**别名：** 十姊妹　　　　　　　　　　　　**科属：** 蔷薇科蔷薇属

**形态特征：** 多花蔷薇的变种。落叶或半常绿灌木，茎直立或攀缘，通常无毛，有皮刺。奇数羽状复叶互生，小叶5～9枚，小叶有锯齿，叶面无毛，叶背有柔毛，托叶篦齿状。花重瓣，深粉红色，伞房花序，常7～10朵簇生，花径约2 cm，具芳香。果近球形，直径6～8 mm，红褐色，有光泽。

**习性、应用及产地分布：** 喜阳光，耐寒、耐旱、耐水湿，适应性强，对土壤要求不严。庭院造景时可布置成花柱、花架、花廊等造型，开花时远看锦绣一片，鲜红艳丽，非常美丽，是优良的护坡及棚架垂直绿化材料。适生于中国长江以北至黄河流域。

**Description:** The variant of floribunda rose, erect or climbing, glabrous, spinescent. Leaves pinnate alternate; leaflets 5–9, serrate, adaxially glabrous, abaxially tomentose. Double petal, dark-pink, corymbs, 7–10 flowers fasciculate, ca. 2 cm in daim., fragrant. Fruit subglobose, 6–8 mm in diam., reddish brown, shiny.

# 木香 | *Rosa banksiae*

**科属：** 蔷薇科蔷薇属

**形态特征：** 落叶或半常绿攀缘灌木，高可达6 m。枝细长，具皮刺（经栽培后有时无刺）。复叶具小叶3～5（7）枚，卵状长椭圆形至披针形，长2～5 cm；叶面有光泽，边缘有细锐齿。顶生伞形花序，花梗细长；花白色、黄色，径约2.5 cm，半重瓣或重瓣，浓香。果近球形，红色，径3～4 mm。花期4—7月，果期8—10月。

**习性、应用及产地分布：** 阳性树种。可耐-5℃左右的低温，忌积水；生长迅速，萌芽力强，耐修剪整形。重要的园林藤本植物，宜设棚架、花廊及凉亭等。产于中国西南及陕西、甘肃、山东等地。

**常见变种及栽培种：**

（1）单瓣白木香（*Rosa banksiae* var. *normalis*），花单瓣，白色，味香；小叶3～7枚。

（2）重瓣白木香（*Rosa banksiae* 'Alba-plena'），花重瓣，白色，香味浓烈；小叶常为3枚。应用广泛。

（3）黄木香（*Rosa banksiae* 'Lutescens'），花单瓣，黄色，无香气。

（4）重瓣黄木香（*Rosa banksiae* 'Lutea'），花重瓣，淡黄色，香味淡；小叶常5枚。

**Description:** Shrubs, climbing, 6 m tall. Branches slender and thorny. Leaves pinnate alternate, leaflets 3–5 (7), elliptic-ovate or lanceolate, 2–5 cm long; adaxially shiny, margin serrate, umbels terminal, flower white or yellow, ca. 2.5 cm in diam., double or semi-double, fragrant. Fruit subglobose, red. Fl. Apr–Jul, fr. Aug–Oct.

## 山木香 | *Rosa cymosa*

**别名:** 小果蔷薇　　　　　　　　**科属:** 蔷薇科蔷薇属

**形态特征:** 常绿攀缘灌木;枝条细长,具较多钩刺。奇数羽状复叶,小叶3～7枚,卵状披针形或长椭圆形,长1.5～5 cm,缘具细锯齿,无毛,叶轴背面具倒钩刺,托叶线形。复伞房花序,小花白色,径2 cm,单瓣,芳香,萼片常羽状裂;花托、花梗有柔毛。花期4—5月。

**习性、应用及产地分布:** 喜温暖、湿润气候及微酸性土壤,耐半荫;萌芽力强,耐修剪。宜在花架、凉廊、墙隅、假山处配置。产于中国华东、中南及西南地区,山野习见。

**Description:** Evergreen shrubs, climbing, with prickles. Leaves odd-pinnate, leaflets 3-7, elliptic-ovate or lanceolate, 1.5-5 cm, margin serrate, glabrous, abaxially, stipules. Complex cymes, flower white, 2 cm in diam., single-layered petal, fragrant, sepals pinnate; stalks, peduncle tomentose. Fl. Apr-May.

## 棣棠 | *Kerria japonica*

**科属：** 蔷薇科棣棠花属

**形态特征：** 落叶丛生无刺灌木，高1.5～2 m。小枝绿色光滑，常拱垂。单叶互生，叶卵形至卵状椭圆形，长4～8 cm，宽2～4.5 cm，先端长尾尖，基部圆形，边缘具尖锐重锯齿，叶背微被短柔毛；叶柄长5～10 mm；托叶与叶柄分离。花单生于侧枝端，金黄色，径3～4.5 cm。瘦果扁球形，黑色，生于盘状花托上。花期4—6月，果期7—8月。

**习性、应用及产地分布：** 亚热带阴性树种。喜温暖、半荫略湿环境，不耐严寒。本种枝叶青翠，花色金黄，是美丽的观赏花木。产于中国和日本，中国黄河流域至华南、西南均有分布。

**Description:** Shrubs, deciduous, 1.5–2 m tall. Branchlets green, usually arcuate. Leaf blade triangular-ovate or ovate, 4–8 cm × 2–4.5 cm, apex acuminate, base rounded, margin sharply doubly serrate, abaxially pilose, petiole 5–10 mm. Flowers golden, 3–4.5 cm in diam. Achenes obovoid, black. Fl. Apr–Jun, fr. Jun–Aug.

# 插田泡 | *Rubus coreanus*

**科属：** 蔷薇科悬钩子属

**形态特征：** 灌木，高1～3m。枝红褐色，被白粉，具钩状扁平皮刺。小叶3～5枚，卵形、菱状卵形或宽卵形，长3～8cm，先端急尖，基部楔形或近圆，叶面无毛或沿叶脉有短柔毛，叶背疏被柔毛或沿叶脉被短柔毛，有不整齐粗锯齿或缺刻状粗锯齿，顶生小叶顶端有时有3浅裂；叶柄长2～5cm，顶生小叶与叶轴均被柔毛和疏生钩状小皮刺。伞房花序顶生，具花数朵；花径0.7～1cm；花萼被灰白色短柔毛，萼片长卵形或卵状披针形；花瓣倒卵形，淡红至深红色；雌蕊多数；子房疏被短柔毛。果近球形，径5～8mm，成熟时深红至紫黑色。花期4—6月，果期6—8月。

**习性及产地分布：** 喜光，忌炎热，适应性较强，一般土壤均可栽种。产于中国西北、西南、东南各省。朝鲜和日本亦有分布。

**Description:** Shrubs, 1–3 m tall. Branchlets reddish brown, with straight flattened curved prickles. Leaves pinnate alternate; leaflets 3–5, ovate, rhombic ovate, or broadly ovate, 3–8 cm long, apex acute, base cuneate to subrounded, adaxially glabrous or pubescent only along veins, abaxially pubescent or only along veins or shortly tomentose, margin irregularly coarse serrate to incised coarsely serrate, sometimes 3-lobed on terminal. Corymbose, several to more than 30-flowered, calyx gray pubescent; petals obovate, pink to dark red, pistils numerous. Fruit subglobose, 5–8 mm in diam., red or purplish black. Fl. Apr–Jun, fr. Jun–Aug.

## 桃 | *Amygdalus persica*

**科属：**蔷薇科桃属

**形态特征：**落叶小乔木，高3～8 m。树皮暗红褐色，粗糙，具皮孔；冬芽常3芽并生，被短柔毛。叶椭圆状披针形，长7～15 cm，宽2～3.5 cm，先端渐尖，基部宽楔形，缘有单细锯齿；叶柄常具腺体。花单生或2～3朵簇生，先叶开放，径2.5～3.5 cm，粉红色，稀白色，近无梗。核果卵球形，径5～7 cm，淡绿白色至橙黄色，表面密被短柔毛，腹缝明显。花期3—4月，果期6—9月。

**习性、应用及产地分布：**温带及亚热带树种。适应性强，喜光，耐寒。多植于庭园观赏。产于中国华中、西北及西南山区。

**Description:** Small trees, 3–8 m tall. Bark dark reddish brown, scabrous, with lenticels; winter buds often 3 in a fascicle. Leaves elliptic-lanceolate, 7–15 cm × 2–3.5 cm, apex acuminate, base broadly cuneate, margin finely to coarsely serrate; petiole often with several nectaries. Flowers solitary, opening before leaves, 2.5–3.5 cm in diam., pink or white, subsessile. Drupe ovoid, 5–7 cm in diam., greenish white to orangish yellow, densely pubescent, ventral suture conspicuous. Fl. Mar–Apr, fr. Jun–Sep.

# 紫叶碧桃 | *Amygdalus persica* 'Atropurpurea'

**科属：** 蔷薇科桃属

**形态特征：** 落叶小乔木，高3～8 m。树皮暗红褐色，老时粗糙呈鳞片状；小枝细长，无毛，有光泽，具皮孔；冬芽常2～3个簇生，中间为叶芽，两侧为花芽。叶片紫红色，长圆披针形、椭圆披针形或倒卵状披针形，长7～15 cm，宽2～3.5 cm，先端渐尖，基部宽楔形，表面光滑无毛，缘具细锯齿。花单生，先于叶开放，直径2.5～3.5 cm；花近无梗；花粉红色，罕为白色。花期4月，果期8—9月。

**习性与应用：** 喜光，耐旱，喜肥沃排水良好土壤，不耐水湿，耐寒。在北京以北地区可露地越冬。因其着花繁密，栽培简易，故南北园林中都有广泛应用。

**Description:** Small trees, 3–8 m tall. Bark dark reddish brown, scabrous; branchlets, slender, glabrous, lustrous, with many small lenticels; winter buds often 2 or 3 in a fascicle. Leaves purple, blade oblong-lanceolate, elliptic-lanceolate, or obovate-oblanceolate, 7–15 cm × 2–3.5 cm, apex acuminate, base broadly cuneate, margin finely to coarsely serrate. Flowers solitary, opening before leaves, 2.5–3.5 cm in diam., petals pink or rarely white, subsessile. Fl. Apr, fr. Aug–Sep.

# 杏 | *Armeniaca vulgaris*

**科属：**蔷薇科杏属

**形态特征：**落叶乔木，高6～12 m。1年生枝灰褐色至红褐色；叶卵形至近圆形，长5～9 cm，宽4～8 cm，先端渐尖，基部圆形，边缘有圆钝锯齿，两面无毛或仅中脉疏被柔毛；叶柄红色，常具1～6个腺体，长2～3.5 cm。花单生，径2～3 cm，白色或稍带粉红色；花萼裂片卵形或椭圆形，花后反折；花瓣5枚，圆形或倒卵形；雄蕊多数；心皮1枚，子房上位。核果卵圆形，径2.5～3 cm，黄白色或黄色，微被短柔毛；核扁平，面光滑。花期3—5月，果期6—7月。

**习性、应用及产地分布：**温带及亚热带阳性树种。能耐严寒，不耐涝，是中国北方著名的早春观赏花木，既可采果又能赏花，园林应用甚多，素有"南梅北杏"之说。可配植于庭前、墙隅、路旁或片植于山坡用于防风固沙。中国秦岭、淮河以北均有分布，尤以黄河流域各省最为集中；日本和朝鲜亦有分布。

**Description:** Trees, deciduous, 6–12 m tall. Bark grayish brown to reddish brown. Leaf blade broadly ovate to orbicular-ovate, 5–9 cm × 4–8 cm, apex acuminate, base orbicular, margin crenate, both surfaces glabrous; petiole red, with 1–6 glands, 2–3.5 cm long. Flowers solitary, 2–3 cm in diam.; petals white or pink, sepals ovate to ovate-oblong, reflexed after anthesis; petals 5, orbicular to obovate. Drupe ovoid, 2.5–3 cm in diam., yellow-white or yellow, pubescent. Fl. Mar–May, fr. Jun–Jul.

# 宫粉梅 | *Armeniaca mume* var. *alphandii*

**科属：** 蔷薇科杏属

**形态特征：** 属梅花品系中真梅系直枝梅类宫粉型。落叶小乔木，高 4～10 m。树皮灰褐色；1年生小枝红褐色，光滑。叶卵形或椭圆形，长 4～8 cm，宽 2.5～5 cm，先端尾尖，基部宽楔形或圆形，两面无毛，边缘具细锐锯齿；叶柄常有腺体。花重瓣，粉红色，先叶开放，径 2.5～5 cm；花萼紫红色。果近球形，核表面具蜂窝状孔穴。花期3月，果期4—6月。

**习性、应用及产地分布：** 亚热带阳性树种。喜温暖、湿润气候，对土壤要求不严，萌芽力强，极耐修剪。宫粉梅是观赏型花梅，开花繁密，花色淡红，尤其难得的是能散发出较为浓郁的清香。早春开花，香色俱佳，是中国著名的观赏花木。产于中国西南地区，长江流域及其以南地区多露地栽植。

**Description:** Small trees, 4-10 m tall. Bark grayish brown, one-year bark reddish brown, smooth. Leaves ovate or elliptic, 4-8 cm × 2.5-5 cm, apex acuminate, base broadly cuneate or subrounded, both surfaces glabrous, margin finely to coarsely serrate; petiole with nectaries. Flowers, pink, opening before leaves, 2.5-5 cm in diam., sepals red-purple. Fruit globose. Fl. Mar-May, fr. Jun-Jul.

# 李 | *Prunus salicina*

**科属：** 蔷薇科李属

**形态特征：** 落叶乔木，高9～12 m。树冠广圆形；老枝紫褐色，小枝红褐色。叶倒卵形或椭圆状倒卵形，长6～12 cm，宽3～5 cm，先端渐尖或短尾尖，基部楔形，叶背无毛或微被柔毛，缘具不规则细钝齿；叶柄顶端有2个腺体或无。花常3朵簇生，白色，径1.5～2.2 cm；萼筒钟状，裂片卵形；雄蕊多数，约与花瓣等长。核果卵球形，径3.5～5 cm，黄色或红色，外被蜡粉；核有沟纹。花期3—4月，果期7—9月。

**习性、应用及产地分布：** 温带阳性树种。对气候适应性强，耐寒，不耐长期积水；花洁白繁密，虽无桃花的娇媚，但以香雅著称。宜植于云雾缭绕的山泉、石林等清净之地，远观更佳。产于中国东北、华北、华东地区，世界各地均有栽培。

**Description:** Trees, deciduous, 9−12 m tall. Branches purplish brown when old and reddish brown when young. Leaf blade obovate or elliptic-obovate, 6–12 cm × 3−5 cm, narrowly elliptic, or rarely oblong-ovate, apex acuminate or short caudate apex, base broadly cuneate, adaxially glabrous or pubescent, margin irregular finely serrate; petiole terminal with two nectaries or not. Three flowers fasciculate, white, 1.5−2.2 cm in diam.; calyx campanulate, lobulate ovate; stamens numerous. Drupe ovoid, 3.5−5 cm in diam., yellow or red, covered with wax and powder; core has grooves. Fl. Mar−Apr, fr. Jul−Sep.

## 紫叶李 | *Prunus cerasifera* f. *atropurpurea*

**别名**：红叶李　　　　　　　　　　　**科属**：蔷薇科李属

**形态特征**：落叶乔木，高4m。小枝暗红色，无毛。叶紫红色，卵形或卵状椭圆形，长3～4.5 cm，基部楔形或近圆形，边缘有圆钝锯齿。花较小，淡粉红色，通常单生，叶前开花或与叶同放。果小，径1～3 cm，暗红色。花期4月，果期8月。

**习性、应用及产地分布**：紫叶李是樱李（*Prunus cerasifera*）的变型，整个生长季节都为紫红色，是园林中常用的彩叶树种；宜植于建筑物前及园路旁或草坪角隅处。

**Description:** Deciduous trees, 4 m tall. Branchlets dark red, glabrous. Leaves purplish red, ovate or oval-elliptic, 3–4.5 cm long, base cuneate or suborbicular, margin obtuse serrate. Flowers smaller, pale pink, usually solitary, flowering before or meanwhile with leaf. Fruit small, 1–3 cm in diam., dark red. Fl. Apr, fr. Aug.

# 杜梨 | *Pyrus betulaefolia*

**科属：** 蔷薇科梨属

**形态特征：** 落叶乔木，高10 m。常具枝刺，小枝幼时密被灰白色绒毛，2年生枝具稀疏绒毛或近无毛。叶菱状卵形至长圆形，长4～8 cm，宽2.5～3.5 cm，先端渐尖，基部宽楔形，幼叶上下密被灰白色绒毛，边缘具尖锐锯齿。花白色，10～15朵成顶生伞形总状花序，花序梗及花梗被毛；花瓣宽卵形；雄蕊20枚，花药紫红色；花柱2～3枚。果小，近球形，径0.5～1 cm，褐色，有淡色斑点，花萼脱落。花期4月，果期8—9月。

**习性、应用及产地分布：** 温带及亚热带阳性树种。适生性强，耐寒，极耐干旱瘠薄；寿命长。树形优美，花色洁白，性强健，对水肥要求也不严，在北方盐碱地区应用较广，可用作防护林、水土保持林或街道、庭院及公园绿化。中国分布较广，华北、西北、长江中下游流域及东北南部地区均有野生分布或栽培。

**Description:** Deciduous trees, 10 m tall. Branches often with thorns, small branches densely grayish white. Leaves rhomboid-ovate or oblong-ovate, 4–8 cm × 2.5–3.5 cm, apex acuminate, base broadly cuneate, young leaves densely grayish white, margin sharply serrate. Flowers white, peduncles and stalks pubescent, pepals broadly ovate; stamens 20, anthers purplish red; styles 2–3. Fruit small, subglobose, 0.5–1 cm in diam., brown, with pale-colored spots, sepals caducous. Fl. Apr, fr. Aug–Sep.

# 厚叶石斑木 | *Rhaphiolepis umbellata*

**别名**：车轮梅　　　　　　　　　　　　**科属**：蔷薇科石斑木属

**形态特征**：常绿灌木，高2～4 m。枝与叶幼时被褐色柔毛，后脱落。叶厚革质，长椭圆形至倒卵形，长4～8 cm，宽2～4 cm，先端圆钝，基部楔形，全缘或有不明显浅齿，叶缘略反卷，叶面深绿色，有光泽；叶柄长5～10 mm。圆锥花序顶生，直立；花白色，倒卵形，径1～1.5 cm；雄蕊20枚，花柱2枚，基部合生。果实球形，径约1 cm，紫黑色带白霜。花期5—6月。

**习性、应用及产地分布**：喜光，耐水湿，耐盐碱，抗风，耐寒，适应性强。耐修剪，形态奇特，可用作庭园及公园绿地的景观植物与绿化树种。产于中国浙江，在日本广泛分布。

**Description:** Evergreen shrubs, 2–4 m tall. Branches and leaves are brown pilose when young, and fall off. Leaves thick leathery, long elliptic to obovate, 4–8 cm × 2–4 cm, apex obtuse, base cuneate, margins entire or with inconspicuously serrate, leaves margin slightly revoled, adaxially dark green, shiny; petiole 5–10 mm. Panicles terminal, erect; flowers white, obovate, 1–1.5 cm in diam.; stamens 20, styles 2, basally connate. Fruit globose, ca. 1 cm in diam., purple-black with hoarfrost. Fl. May–Jun.

# 樱花 | *Cerasus serrulata*

**科属**：蔷薇科樱属

**形态特征**：落叶乔木，高3～8m。树皮灰褐色或灰黑色，有光泽，横裂皮孔。叶卵状椭圆形，互生，长4～10cm，先端渐尖或骤尾尖，缘有芒状单锯齿或重锯齿，叶柄密被柔毛，顶端具1～2个腺体或无。花白色至粉红色，3～4朵成伞形总状花序，先叶开放，径3～3.5cm，花瓣先端凹缺；萼筒管状，花梗与萼筒被柔毛；萼片略短于萼筒，边缘具腺齿。核果近球形。花期4月，果期5月。

**习性、应用及产地分布**：温带阳性树种。喜光、耐寒、忌积水、不耐盐碱。樱花妩媚多姿，既有梅花之幽丽，亦有桃花之粉艳，是美丽的庭园观花树种。产于中国、日本与朝鲜，华东、台湾及两广地区栽培较多。

**Description:** Deciduous trees, 3–8 m tall. Bark grayish brown or grayish black, glossy, with transverse crack lenticels. Leaf blade ovate-elliptic to obovate-elliptic, 4–10 cm long, base rounded, margin acuminately serrate or biserrate and teeth with a minute apical gland, apex acuminate. Inflorescences corymbose-racemose or subumbellate, 3- or 4-flowered, involucral bracts brownish red, obovate-oblong, adaxially villous; peduncle 5–10 mm, glabrous. Pedicel 1.5–2.5 cm, glabrous, sparsely pilose, or pubescent. Petals white or rarely pink, obovate, apex emarginate. Drupe purplish black, globose to ovoid, 8–10 mm in diam. Fl. Apr, fr. May.

# 郁李 | *Cerasus japonica*

**科属**：蔷薇科樱属

**形态特征**：落叶灌木，高 1～1.5 m。小枝灰褐色，无毛。叶卵形至卵状披针形，长 4～7 cm，宽 1.5～2.5 cm，先端长尾状，基部圆形，边缘具尖锐重锯齿；托叶条形。花 1～3 朵簇生，花叶同放或先叶开放，径 1～1.5 cm。花萼在花后反折，花瓣粉红色或近白色，倒卵形。核果近球形，径约 1 cm，果成熟深红色，光滑有光泽。花期 4—5 月，果期 7—8 月。

**习性、应用及产地分布**：温带阳性树种。性强健，耐严寒，抗旱性强，适宜于石灰岩山地和肥沃湿润的砂质壤土，一般土壤均可栽培。根蘖多，簇生成丛。花开时繁花压树，灿若云霞；果熟时，丹实满枝，是花果同赏的园林树种，宜丛植、群植于草坪、山石旁、林缘、建筑物前，点缀于庭院、屋旁或路边。原产于中国东北与华东地区。

**Description:** Deciduous shrubs, 1–1.5 m tall. Branchlets grayish brown, glabrous. Leaf blade ovate to ovate-lanceolate, 4–7 cm × 1.5–2.5 cm, abaxially pale green and glabrous or pilose along veins, adaxially dark green and glabrous, base rounded, margin acutely incised biserrate or deeply serrate, apex acuminate. Sepals elliptic, slightly longer than hypanthium, reflexed; petals pink or white, obovate. Drupe dark red, subglobose, ca. 1 cm in diam, endocarp smooth. Fl. Apr–May, fr. Jul–Aug.

# 31. 豆科

## 合欢 | *Albizia julibrissin*

**别名**：夜合花　　　　　　　　　**科属**：豆科合欢属

**形态特征**：落叶中乔木，高 15 m。树冠宽广而平展；树皮灰褐色，不裂或浅纵裂；小枝褐绿色。2回羽状复叶，复叶具羽片4～12对，1羽片有小叶10～30对，夜间闭合，小叶镰刀状长圆形，长6～12 mm，宽1～4 mm，全缘，中脉明显偏向上缘。头状花序侧生于叶腋或顶生，花序梗细长。花萼、花瓣黄绿色，微香；花丝细柔，粉色，基部合生，长2.5～3 cm，伸出花冠筒外，如绒缨状。荚果带状，长8～17 cm，宽1.2～2.5 cm，老荚无毛；种子8～14粒。花期6—7月，果期9—10月。

**习性、应用及产地分布**：温带、亚热带及热带阳性树种。适生于温暖、干燥气候，不耐严寒，耐瘠薄，不耐水涝；生长快，浅根性，不耐修剪；抗污染能力强。树形开展，绿荫如伞，叶纤细如羽，盛夏时红色的绒花开满树，是优良的庭园观赏树种，适宜种植于林缘、草坪、山坡、池畔、水滨、河岸和溪旁。分布范围极广，产于中国黄河、长江及珠江流域各省。

**Description:** Deciduous trees, 15 m tall. Crown open. Stipules deciduous, linear-lanceolate, smaller than leaflets, pinnae 4–12 pairs; leaflets 10–30 pairs, obliquely linear to oblong, 6–12 mm × 1–4 mm, main vein close to upper margin, base truncate. Flowers pink. Corolla ca. 8 mm, filaments pink. Legume strap-shaped, flat, glabrous. Fl. Jun–Jul, fr. Sep–Oct.

## 银荆 | *Acacia dealbata*

**科属：** 豆科金合欢属

**形态特征：** 常绿乔木，高可达25 m。树干较直，树皮银灰色；小枝被绒毛。2回羽状复叶互生，羽片8～20对，小叶极小，30～40对，线形，长2～4 mm，两面有毛，银灰色。总叶轴上每对羽片间有1腺体。头状花序，球形，具小花30～40朵，芳香。荚果长带形，无毛。种子卵圆形，黑色，有光泽。花期1—4月，果期7—8月。

**习性、应用及产地分布：** 喜光，不耐荫，喜温暖、湿润气候。生长迅速，抗逆性强，适作荒山绿化先锋树及水土保持树种，可作行道树或在庭园作孤植、丛植布置。原产于澳大利亚，中国云南、广西、福建有引种。

**Description:** Evergreen trees, 25 m tall. Young branchlets slightly angular with ridges gray tomentose, glaucous. Leaves argenteous to greenish or golden, rachis not angulate, glands at rachis of pinna insertion, pinnae 8–20; leaflets 30–40 pairs, linear, 2–4 mm long, abaxially or both surfaces gray-white pubescent. Flowers yellowish or orange-yellow. Legume red-brown or black, glabrous. Seeds elliptic, black, flat. Fl. Jan–Apr, fr. Jul–Aug.

## 云实 | *Caesalpinia decapetala*

**科属：** 豆科云实属

**形态特征：** 落叶攀缘灌木，树皮暗红色。枝、叶轴和花序梗密被柔毛和倒钩刺。2回羽状复叶，羽片3～8对，每羽片具小叶6～12对，长椭圆形，长1～2.5 cm，宽6～8 mm，两端近圆钝。顶生总状花序长15～30 cm，花黄色，盛开时反卷；雄蕊与花瓣近等长。荚果长椭圆形，长6～12 cm，宽2.5～3 cm，沿腹缝线具狭翅，开裂；种子6～9粒。花期5月，果期8—10月。

**习性、应用及产地分布：** 热带及亚热带阳性树种。不耐寒，对土壤要求不严，耐瘠薄，适宜于石灰岩发育的山地黄壤。生长快，萌蘖性强。春花繁盛、夏果低垂，常作篱垣或花架和花廊垂直绿化，也可丛植于路边、林缘和旷地。产于中国华北、秦岭以南至华南、西南地区山坡、丘陵灌丛中及平原、河旁。

**Description:** Deciduous climbing shrubs, with copious prickles, bark dull red. Branches, rachis of leaves, and inflorescence with recurved prickles and pubescent. Bipinnate leaves 20–30 cm, pinnae 3–8 pairs; leaflets 6–12 pairs, oblong, 1–2.5 cm × 6–8 mm, both surfaces puberulent. Racemes terminal, 15–30 cm, with abundant flowers; petals reflexed at anthesis, yellow; stamens sub-equal to petals in length. Legume chestnut-brown, shiny, oblong-ligulate, 6–12 cm × 2.5–3 cm, apex prolonged into a sharp beak, dehiscence along abdominal of legume. Seeds 6–9. Fl. May, fr. Aug–Oct.

# 紫荆 | *Cercis chinensis*

**别名：** 满条红、老茎生花  **科属：** 豆科紫荆属

**形态特征：** 落叶灌木或小乔木，高2～6 m，栽培时常呈丛生灌木状。树皮灰白色，老时粗糙纵裂。叶心形，长6～14 cm，宽5～14 cm，先端急尖，基部心形，上面光泽，下面微被毛或白粉，全缘。花4～10朵簇生于老枝、主干上，无花序梗，先叶开放；花假蝶形，紫红色，龙骨瓣基部具深紫色斑纹，花瓣长1～1.4 cm。荚果薄革质，带状狭披针形，顶端稍弯，长5～10 cm，宽1.3～1.5 cm，沿腹缝线具窄翅，成熟时不开裂；种子2～8粒，扁圆形，近黑色。花期3—4月，果期8—10月。

**习性、应用及产地分布：** 亚热带阳性树种。喜温热、湿润气候，不耐水淹；萌蘖性强，耐修剪。早春繁花簇生于枝间，满树紫红，鲜艳夺目，是良好的庭园观花树种；因多干簇生，不易单植。产于中国黄河流域及其以南各地。

**Description:** Shrubs or small trees, deciduous, 2–6 m tall. Bark and branchlets grayish white. Leaf blade suborbicular or triangular-orbicular, 6–14 cm × 5–14 cm, papery, both surfaces usually glabrous, or abaxially puberulent on veins, base shallowly to deeply cordate. Flowers red or pink, 4–10 clustered on branches or trunk. Legume greenish, compressed, narrowly oblong, 5–10 cm × 1.5 cm. Seeds 2–8, blackish brown, shiny, broadly oblong. Fl. Mar–Apr, fr. Aug–Oct.

# 紫叶加拿大紫荆 | *Cercis canadensis* 'Purpurea'

**科属：** 豆科紫荆属

**形态特征：** 落叶灌木或小乔木，高达12 m。叶广卵状心形，叶片紫红色。花假蝶形，淡玫瑰红色，长约1.3 cm；4～6朵簇生。

**产地分布：** 此种是加拿大紫荆的栽培品种，是良好的彩叶树种，用于庭园观赏。原产于加拿大南部及美国东部，中国上海、杭州、武汉等地有栽培。

**Description:** Shrubs or small trees, deciduous, 12 m tall. Leaf blade suborbicular or triangular-orbicular, base shallowly to deeply cordate, purplish red. Flowers rose-pink, 4-6 clustered. This is a cultivated variety of Canadian redbud, which is a good colored-leaf species, used for garden appreciation. Native to southern Canada and eastern United States. Cultivated in Shanghai, Hangzhou, Wuhan and other places in China.

# 皂荚 | *Gleditsia sinensis*

**别名：** 皂角　　　　　　　　　　　　**科属：** 豆科皂荚属

**形态特征：** 落叶乔木，高可达15～20 m。树皮灰黑色，浅纵裂；枝干常具粗壮的圆锥状分枝刺。1回羽状复叶有小叶6～14对，小叶长卵形，长3～8 cm，宽1～4 cm，先端钝圆，基部圆形，边缘有细齿；下面中脉两侧及叶柄被白色短柔毛。花杂性；总状花序腋生或顶生，长5～14 cm，被柔毛；花瓣4枚，黄白色，被绒毛。荚果带状，长12～30 cm，平直不扭曲，果肉稍厚；种子多数，光亮。花期3—5月，果期5—10月。

**习性、应用及产地分布：** 温带及亚热带阳性树种。适生于温暖、湿润气候及深厚肥沃的土壤，在石灰质及盐碱土甚至黏土或砂土上均能正常生长。深根性，生长速度慢，寿命长。冠大荫浓，适宜作庭荫树及荒山造林树种。产于中国黄河流域以至华南、西南，多生于平原、山谷及丘陵地区。

**Description:** Trees, deciduous, 15–20 m tall. Branches grayish to deep brown. Spines robust, terete, conical, to 16 cm, often branched. Leaves pinnate; leaflets 6–14 pairs, ovate-lanceolate to oblong, 3–8 cm × 1–4 cm, papery, abaxially slightly pubescent on midvein, adaxially puberulent; reticulate veinlets conspicuously raised on both surfaces. Flowers polygamous, yellowish white, in axillary or terminal, puberulent racemes 5–14 cm. Legume curved, strap-shaped, 12–30 cm × 2–4 cm, straight or twisted, with slightly thick pulp. Seeds numerous, brown, shiny. Fl. Mar–May, fr. May–Oct.

# 山皂荚 | *Gleditsia japonica*

**别名：** 日本皂荚　　　　**科属：** 豆科皂荚属

**形态特征：** 落叶乔木，高可达14 m。枝刺多密集，基部扁圆；小枝淡紫色。1回兼有2回偶数羽状复叶，常簇生，小叶6～11对，卵状长椭圆形，长2～8.5 cm，先端钝尖或微凹，基部稍偏斜，叶背中脉被柔毛，边缘有细锯齿。花杂性异株，黄白色，总状花序腋生或顶生，长5～14 cm，被柔毛。荚果革质，常不规则扭转，长12～37 cm，宽2～4 cm，棕黑色，果肉肥厚，种子多数。花期5—6月，果期6—10月。

**习性、应用及产地分布：** 喜光、耐寒、耐干旱，喜肥沃深厚土壤。应用同皂荚。产于中国华北至华东地区，日本、朝鲜也有分布。

**Description:** Trees, deciduous, 14 m tall. Spines slightly flat, robust, often branched. Leaves pinnate or bipinnate; leaflets 6–11 pairs, ovate-oblong, 2–8.5 cm long, base slightly oblique, apex rounded, sometimes emarginate. Flowers yellowish white, in axillary or terminal, puberulent spikes. Legume compressed, strap-shaped, 12–37 cm × 2–4 cm, irregularly twisted. Seeds numerous. Fl. May–Jun, fr. Jun–Oct.

# 美国金叶皂荚 | *Gleditsia triacanthos* 'Sunburst'

**科属：** 豆科皂荚属

**形态特征：** 落叶乔木，高可达30～45 m。1～2回羽状复叶，常簇生，小叶5～16对，卵状披针形，长2～3.5 cm，叶面有光泽，叶背中脉有毛，幼叶金黄色，后渐变绿色。荚果镰刀形，长30～45 cm，熟时褐色，疏生灰黄色柔毛。

**习性、应用及产地分布：** 喜深厚、肥沃而排水良好土壤。适作园林观赏树、庭荫树。原产于美国。

**Description:** Trees, deciduous, up to 30–45 m tall. Leaves pinnate or bipinnate; leaflets 5–16 pairs, adaxially shiny, abaxially puberulent on midvein, leaves golden yellow when young. Legume 30–45 cm long, brown when mature, with grayish yellow pubescence.

# 龙牙花 | *Erythrina corallodendron*

**别名：** 象牙红　　　　　　　　　　**科属：** 豆科刺桐属

**形态特征：** 落叶灌木或小乔木，高可达 3～6 m。枝干散生皮刺。三出复叶，顶生小叶较侧生小叶大，菱状卵形，先端渐尖，全缘；叶柄和叶背中脉具刺。总状花序长 30 cm 以上；花萼钟状，红色；花冠深红色，长 4.5～6 cm，狭而近于闭合，先叶开放或与叶同放；雄蕊 2 体；子房具长柄，被白色短柔毛。荚果圆柱形，长约 10 cm；种子多数，深红色，有黑斑。花期 6—7 月，果期 8—9 月。

**习性、应用及产地分布：** 热带阳性树种。喜高温、多湿环境，不耐寒。初夏开花，深红色的总状花序，艳丽夺目。是著名的观赏花木，适于公园和庭院栽植。原产于美洲的热带地区，中国华南各地至台湾有栽培。

**Description:** Shrubs or small trees, deciduous, 3–6 m tall. Branches with prickles. Leaves pinnately 3-foliolate, terminal lobule larger than lateral lobule, rhomboid ovate, apex acuminate, margin entire. Raceme longer than 30 cm, calyx campanulate red, corolla deep red, 4.5–6 cm long; diadelphous stamen 2 bodies; ovary long stipitate, white pubescent. Legume cylindrical, 10 cm long, seeds numerous dark red. Fl. Jun–Jul, fr. Aug–Sep.

# 紫穗槐 | *Amorpha fruticosa*

**科属：** 豆科紫穗槐属

**形态特征：** 丛生灌木，高1～4 m。树皮暗灰色，平滑。羽状复叶具小叶11～25枚，小叶卵形，长2～4 cm，先端圆或微凹，有芒尖，叶背密被白色柔毛。穗状花序1至数个顶生，长7～15 cm，小花暗紫色；雄蕊10枚，花药黄色。荚果短镰形，种子1粒。花期5—6月，果期9—10月。

**习性、应用及产地分布：** 温带及亚热带阳性树种。喜干冷气候，耐寒性强，耐旱、耐水湿，生长迅速，萌芽性强，侧根发达，抗风沙。是优良的水土保持、防护林树种，可作城市道路隔离带树种。原产于美国，广布于中国东北、华北至长江流域各地。

**Description:** Deciduous shrubs, 1–4 m tall. Stems pubescent, glabrescent. Leaflets 11–25, ovate to elliptic, 2–4 cm long, abaxially white puberulent, adaxially glabrous or sparsely pubescent. Racemes 1 to many, terminal or subterminal, 7–15 cm long, densely pubescent; stamens 10, anthers yellow. Legume dark brown, oblong, curved, 1 seed. Fl. May–Jun, fr. Sep–Oct.

## 紫藤 | *Wisteria sinensis*

**科属：** 豆科紫藤属

**形态特征：** 落叶大藤本植物，茎较粗壮，长可达18～30 m。羽状复叶互生，具小叶7～13枚，小叶纸质，卵形至卵状披针形，长5～8 cm，顶端小叶最大，基部1对最小。花数朵，成总状花序着生于去年生短枝顶端，长15～30 cm；花冠紫色或紫红色，芳香，旗瓣花开后反折，龙骨瓣较翼瓣短；子房密被柔毛，胚珠6～8个。荚果扁，木质，长10～15 cm，密生灰黄色茸毛，开裂；种子1～5粒。花期4—5月，果期5—8月。

**习性、应用及产地分布：** 亚热带及温带树种。喜光，略耐荫，较耐寒；生长迅速，主根深，寿命长。庭院棚架植物，已有2 000多年栽培历史，绿蔓浓荫，春天先花后叶，繁花满架，十分雅致，适宜攀缘于花架、门廊、墙壁、枯树旁及叠石、山坡。产于中国黄河流域以南至华南北部、西南，山林中有野生紫藤。

**Description:** Lianas to 18–30 m long. Pinnate leaves with leaflets 7–13, blades elliptic-ovate to lanceolate-ovate, 5–8 cm long with basal pair smallest and becoming larger apically. Racemes terminal or axillary from branchlets of previous year, 15–30 cm long; corolla purple or occasionally white, standard orbicular, sometimes retuse, glabrous, apex truncate; ovary tomentose, with 6–8 ovules. Legume oblanceolate, 10–15 cm long, tomentose, hanging on branches persistently. Seeds 1–5 per legume. Fl. Apr–May, fr. May–Aug.

# 常春油麻藤 | *Mucuna sempervirens*

**别名：** 常绿油麻藤　　　　　　　　**科属：** 豆科油麻藤属

**形态特征：** 常绿藤木植物，长达10 m以上。三出复叶互生，薄革质而有光泽；顶生小叶卵状椭圆形，长7～12 cm，两面无毛；侧生小叶基部偏斜呈斜卵形，叶柄膨大。总状花序生于老茎上，长10～36 cm，下垂；花大，暗紫色，蜡质。荚果长条状，长约40 cm，被红褐色短伏毛和长刚毛，种子间缢缩。花期4—5月。

**习性、应用及产地分布：** 耐荫，喜温暖、湿润气候，耐旱，是美丽的棚荫及垂直绿化材料。产于中国西南至东南地区，日本也有分布。

**Description:** Lianas, evergreen, more than 10 m long. Ternately compound leaf, alternate, leathery and lustrous; terminal lobules ovate-elliptic, 7–12 cm long, glabrous both sides; lateral lobules obaxially ovate at base; petiole enlarged. Racemes on old stems, 10–36 cm long, pendulous; flowers large, dark purple, waxy. Pod elongate, ca. 40 cm long, covered with reddish-brown short pubescent and long setae. Fl. Apr–May.

# 龙爪槐 | *Sophora japonica* f. *Pendula*

**科属：** 豆科槐属

**形态特征：** 落叶乔木；枝条扭转下垂，树冠伞形；树皮暗灰色；小枝绿色。羽状复叶具小叶7～19对；小叶对生，卵圆形，长2.5～7.5 cm，宽1.5～3 cm，叶背苍白色，疏生短柔毛，叶轴基部膨大。花冠蝶形，黄白色，旗瓣具紫色脉纹。荚果肉质，不开裂。

**习性与应用：** 此为人工嫁接品种，常用于庭园门旁对植或路边列植观赏。

**Description:** Deciduous trees; branches torsion pendulous, canopy umbrella-shaped; bark dark gray; branchlets green. Pinnate leaves with leaflets 7–19 pairs; leaflets opposite, ovoid, 2.5–7.5 cm × 1.5–3 cm, pale back, sparsely pubescent, base of leaf axis dilated. Corolla shaped like a butterfly, yellow and white, vexilla with purple veins. Pod fleshy, not dehiscent. Often cultivated for ornamental purposes near gates or roadsides.

# 蝴蝶槐 | *Sophora japonica* var. *oligophylla*

**别名**：畸叶槐、五叶槐　　　　　**科属**：豆科槐属

**形态特征**：小叶5～7枚，簇集在一起，大小和形状均不整齐，顶生小叶常3裂，侧生小叶下部常具大裂片，叶下面被毛。圆锥花序生于枝端。

**习性、应用及产地分布**：阳性树种，生长势较弱。叶形奇特，常用于庭园观赏。中国山东、河北、河南及辽宁南部有栽培。

**Description:** Leaflets 5-7, clustered together, size and shape irregular, terminal lobules often 3-lobed, side lobules often with large lobes, abaxially puberulent. Panicles at the end of branches. Leaf shape peculiar, often used in the garden. Cultivated in Shandong, Hebei, Henan and the southern part of Liaoning Province, China.

## 刺槐 | *Robinia pseudoacacia*

**别名：** 洋槐　　　　　　　　　　　　　　**科属：** 豆科刺槐属

**形态特征：** 落叶乔木，高10～20 m。树皮灰黑褐色，深纵裂，枝具托叶刺。羽状复叶具小叶7～19对；小叶卵状长圆形，长2～5 cm，全缘，先端微凹并有小刺尖。总状花序长10～20 cm，下垂；花冠白色，芳香。荚果褐色，扁平光滑，长5～12 cm；种子3～10粒。花期4—6月，果期8—9月。

**习性、应用及产地分布：** 温带强阳性树种。喜干爽、冷凉气候，较抗旱，不耐水湿；对土壤要求不严；萌蘖力强，生长快，浅根性，侧根发达，抗风能力弱。树冠高大，叶色鲜绿，花季白花绿叶相映，素雅而芳香；具改良土壤、提高地力之效。可作庭荫树、工矿区绿化及水土保持林、荒山绿化的先锋树种。原产于美国东部和中部，现中国南北各地广泛栽培。

**Description:** Trees, deciduous, 10–20 m tall. Bark gray-brown to dark brown, deep longitudinally fissured, stipulate spines up to 2 cm. Leaflets 7–19 pairs; leaflet blade oblong, elliptic or ovate, 2～5 cm long, margin entire, apex rounded, retuse and apiculate. Racemes axillary, 10–20 cm, pendulous, many flowered, fragrant, bracts caducous; corolla white, fragrant. Legume brown, linear-oblong, 5–12 cm long. Seeds 3–10. Fl. Apr–Jun, fr. Aug–Sep.

# 锦鸡儿 | *Caragana sinica*

**科属：** 豆科锦鸡儿属

**形态特征：** 落叶小灌木，高1～2 m。小枝黄褐色，细长有棱。羽状复叶互生，小叶4枚，倒卵形，长1～3.5 cm，宽0.5～1.5 cm，先端微凹有短尖头，无柄；叶轴顶端硬化成针刺。花单生于短枝叶腋；花萼钟状；花冠黄色或深黄色，凋谢时红褐色，长2～3 cm。荚果圆筒状，稍扁，长3～3.5 cm，宽约0.5 cm。花期4—5月，果期7—8月。

**习性、应用及产地分布：** 喜光，喜温暖，耐干旱，可作观花刺篱及岩石园材料。中国华北、华东、华中及西南地区均有分布。

**Description:** Shrubs, deciduous, 1–2 m tall. Branchlets yellow-brown, slender and carinal. Leaves pinnate or sometimes digitate, 4-foliolate; leaflet blade obovate to oblong-obovate, 1–3.5 cm × 0.5–1.5 cm, apical pairs often largest, apex rounded and mucronate. Flowers solitary; pedicel ca. 1 cm, articulate at middle; calyx tube campanulate; corolla yellow or deep yellow, 2–3 cm. Legume cylindric, 3–3.5 cm × 0.5 cm. Fl. Apr–May, fr. Jul–Aug.

# 红豆树 | *Ormosia hosiei*

**科属：** 豆科红豆属

**形态特征：** 常绿乔木，高20～30 m。树皮灰绿色，光滑；小枝绿色。羽状复叶互生，具小叶5～7枚；小叶薄革质，卵形至椭圆形，长5～14 cm，无毛。圆锥花序，长15～20 cm，下垂；花白色或淡紫色，芳香。荚果卵圆形，长1.5～1.8 cm；种子1～2粒，扁圆形，鲜红色，种脐白色。花期4—5月，果期10—11月。

**习性、应用及产地分布：** 适生于温暖、湿润气候，较耐寒；主根明显，根系发达。树冠呈伞状开展，冠大荫浓，树姿清秀，枝叶翠绿，可植为庭荫树、行道树或片林。产于中国陕西、甘肃东南部及长江中下游地区，是本属中分布于纬度最北地区的种类。国家二级重点保护植物。

**Description:** Trees, evergreen, 20-30 m tall. Bark grayish green, smooth; branchlets green, yellowish-brown pubescent, becoming glabrescent; winter buds brownish yellow pubescent. Leaves pinnate, leaflets 5-7, thinly leathery, ovate to elliptic, 5-14 cm long, glabrous. Panicles terminal or axillary, 15-20 cm, pendulous. Flowers white or pale purple, fragrant. Legumes suborbicular, compressed, 1.5-1.8 cm long. Seeds 1-2, red, suborbicular or elliptic. Fl. Apr-May, fr. Oct-Nov.

# 胡枝子 | *Lespedeza bicolor*

**科属：** 豆科胡枝子属

**形态特征：** 落叶灌木，高 1～3 m，多分枝。三出复叶互生，有长柄；小叶卵状椭圆形，长 1.5～6 cm，宽 1～3.5 cm，先端钝圆或微凹，具短刺尖，基部近圆形或宽楔形，全缘，叶面绿色，无毛，叶背面被疏柔毛，后渐脱落。总状花序腋生；花梗短；花冠紫红色，旗瓣倒卵形，翼瓣较短，龙骨瓣与旗瓣近等长。荚果稍扁，密被短柔毛。花期 7—9 月，果期 9—10 月。

**习性、应用及产地分布：** 喜光，耐半荫，耐寒，耐干旱瘠薄，适应性强。宜作水土保持及防护林下层树种，花美丽，常用于庭园观赏。产自中国东北、内蒙古、华北至长江以南广大地区。

**Description:** Shrubs, deciduous, 1–3 m tall, much branched. Leaves pinnately 3-foliolate; leaflets abaxially pale green, adaxially green, oval to elliptic, 1.5–6 cm × 1–3.5 cm, abaxially pilose, adaxially glabrous, base subrounded or broadly cuneate, apex obtuse-rounded or emarginate, mucronate. Racemes axillary; corolla reddish purple, apex emarginate, wings suboblong, base auriculate, clawed, keel subequal to standard, base long clawed, apex obtuse. Ovary hairy. Legume obliquely obovoid, slightly flat, densely pubescent. Fl. Jul–Sep, fr. Sep–Oct.

# 花木蓝 | *Indigofera kirilowii*

**科属：** 豆科木蓝属

**形态特征：** 落叶灌木，高1～1.5 m。羽状复叶互生，小叶7～11枚，卵状椭圆形至倒卵形，长1.5～3 cm，两面疏生白色丁字毛。腋生总状花序与复叶近等长，花淡紫红色，无毛，长1.5～2 cm。花期5—6月，果期8—9月。

**习性、应用及产地分布：** 耐贫瘠，耐干旱，适应性强，对土壤要求不严。本种枝叶扶疏，花大而美丽，是盛夏良好的观花植物，适宜种于庭园观赏，也可作为山坡覆盖材料。产于中国东北南部、华北至华东地区。

**Description:** Shrubs, deciduous, 1–1.5 m tall. Leaves pinnate, alternate, 7–11 foliolate; leaflet blade opposite, broadly ovate, ovate-rhombic or elliptic, 1.5–3 cm long, both surfaces with appressed medifixed trichomes. Axillary racemes, almost as long as the leaves. Corolla pink or rarely white, standard elliptic, glabrous. Fl. May–Jun, fr. Aug–Sep.

# 32. 大戟科

## 山麻杆 | *Alchornea davidii*

**科属：** 大戟科山麻杆属

**形态特征：** 落叶丛生灌木，高1～5 m。老枝红褐色，幼枝密被灰白色短绒毛。叶广卵形或圆形，长8～15 cm，宽7～14 cm，先端短尖，基部心形，幼叶红色或紫红色，成熟叶绿色，叶背有绒毛，缘有齿；基出脉3条；叶柄长2～10 cm，具2个以上斑状腺体。雌雄异株，雄花密集成短穗状花序，长1.5～3.5 cm；雌花4～7朵成疏散总状花序，长4～8 cm。雄花花萼4片，雄蕊6～8枚，紫色；雌花花萼5片，子房被绒毛，3室，花柱3个。蒴果近球形，径1～1.2 cm，密被短柔毛；种子褐色，具小瘤体。花期3—5月，果期6—7月。

**习性、应用及产地分布：** 喜光，稍耐荫，喜温暖、湿润气候；对土壤要求不严，萌蘖性强。早春嫩叶紫红，鲜艳如花，是良好的观茎、观叶树种，丛植于常绿树种前或草坪边缘，也可在庭院、路边、水滨、山石旁与早春花灌木配植一处，相互映衬。主产于中国长江流域地区。

**Description:** Deciduous shrubs, 1–5 m tall. Branchlets gray tomentulose, becoming puberulent. Leaf blade broadly ovate or subrounded, 8–15 cm × 7–14 cm. Peduncle subsessile, bracts ovate, pubescent. Male flowers 3–5 per bract, bud globose, glabrous, sepals 4, stamens 6–8; female inflorescences terminal, unbranched, pubescent. Ovary subglobose, tomentose. Fl. Mar–May, fr. Jun–Jul.

# 乌桕 | *Sapium sebiferum*

**科属：** 大戟科乌桕属

**形态特征：** 落叶乔木，高18 m。全株含乳汁；树皮暗灰色，有纵裂纹；小枝纤细。叶纸质，菱状广卵形，长3～8 cm，宽3～9 cm，先端尾状，基部宽楔形，全缘；叶柄长2.5～6 cm。复总状花序，长6～12 cm；雄花常生于花序上部，雌花生于花序基部，花柱基部合生，柱头外卷。蒴果梨状球形，木质，径1～1.5 cm，成熟时黑色。花期6—7月，果实11月成熟。

**习性、应用及产地分布：** 喜光，喜温暖气候，畏寒冷；对土壤要求不严，耐水湿；主根发达，抗风力强；生长快，寿命较长。树冠整齐，叶形秀丽，春秋叶色红艳可爱，是优良的园林绿化及观赏树种。宜植于水边、湖畔、山坡和草坪边缘。主产于中国黄河流域以南、长江流域和珠江流域至西南各地。

**Description:** Deciduous trees, up to 18 m tall. Entire plants producing latex; bark dark gray, with longitudinal stripes. Leaf blade rhomboid, rhomboid-ovate, broadly ovate, or rarely rhomboid-obovate, 3–8 cm × 3–9 cm, base broadly rounded, truncate, or sometimes shallowly cordate, margin entire. Flowers yellowish green in terminal 6–12 cm racemes; monoecious, female in lower part, male in upper part or male throughout. Capsules subglobose to pyriform-globose, black when mature. Fl. Jun–Jul, fr. Nov.

# 重阳木 | *Bischofia polycarpa*

**科属：** 大戟科重阳木属

**形态特征：** 落叶乔木，高15 m。树皮褐色，纵裂，具皮孔。三出复叶，顶生小叶通常较两侧的大，叶纸质，卵形，长5～9 cm，先端突尖或渐尖，基部圆形或浅心形，缘有细锯齿，叶柄长9～13 cm。花黄绿色，春季与叶同放；总状花序着生于新枝下部。浆果径5～7 mm，熟时红褐色。花期4—5月，果期10—11月。

**习性、应用及产地分布：** 喜光，稍耐荫；喜温暖、湿润气候，耐寒力弱；生长较快，根系发达，抗风力强；对二氧化硫等有毒气体有一定抗性。树姿优美，枝叶茂盛，早春叶色鲜嫩光亮，入秋叶色转红，颇为美观。南方重要行道树种，同时，宜作庭荫树，在草坪、湖畔、溪边丛植点缀。产于中国秦岭、淮河流域以南至福建、广东北部海拔1 000 m以下山地林中或平原栽培。

**Description:** Trees to 15 m tall, deciduous. Bark brown, longitudinally fissured, with lenticels. Leaves palmately 3-foliolate with petioles 9-13 cm long; leaflet blade ovate or elliptic-ovate, sometimes oblong-ovate, 5–9 cm long, base rounded or shallowly cordate, apex acute or shortly acuminate. Inflorescences yellowish green, pendent racemes, on lower parts of previous year's branches, generally appearing in spring with leaves. Berries, 5–7 mm in diam., brown-red when mature. Fl. Apr–May, fr. Oct.–Nov.

## 33. 芸香科

### 野花椒 | *Zanthoxylum simulans*

**科属**：芸香科花椒属

**形态特征**：落叶灌木或小乔木，高3～8 m。全株均具挥发性香气，茎干具增大的皮刺。奇数羽状复叶互生，具小叶5～9枚，小叶卵形至卵状椭圆形，长1.5～5 cm，叶背中脉及叶轴上常有刺毛，边缘具细钝锯齿；叶轴具窄翅。花小，黄绿色，聚伞状圆锥花序；蓇葖果球形，成熟时红色或紫红色，密生凸起油腺点。花期5—6月，果期9—10月。

**习性、应用及产地分布**：喜阳光，耐干旱。可植于房前屋侧、路边坡地，也可植为绿篱，作隔离之用。产于中国青海、甘肃、山东、河南、安徽、江苏、浙江、湖北、江西、台湾、福建、湖南及贵州东北部。

**Description:** Shrubs or small trees, 3–8 m tall, deciduous. Stems and branchlets with prickles. Leaves 5–9 foliolate, leaflets opposite, ovate, ovate-elliptic, or lanceolate, 1.5–5 cm long, rachis winged. Inflorescences terminal, small, yellowish green; fruit follicles reddish brown, oil glands numerous and slightly protruding. Fl. May–Jun, fr. Sep–Oct.

# 竹叶椒 | *Zanthoxylum armatum*

**别名：** 竹叶花椒    **科属：** 芸香科花椒属

**形态特征：** 落叶灌木或小乔木，高4 m。枝上皮刺对生。奇数羽状复叶互生，小叶3～5枚，卵状披针形，长5～9 cm，边缘小齿下有油腺点；复叶轴具翅和针状皮刺。花黄绿色，成腋生圆锥花序，蓇葖果红色。花期4—5月，果期8—10月。

**习性、应用及产地分布：** 喜光，略耐寒，耐旱，对土适应性强。可用于庭园观赏。分布于中国长江流域至华南、西南各地。

**Description:** Shrubs or small trees, 4 m tall, deciduous. Prickles opposite on branchlets. Odd-pinnate leaves 3–5 foliolate, ovate lanceolate, 5–9 cm long, margin crenate with oil glands; rachis with wings and prickles. Flowers yellow-green, panicle axillary, follicles red. Fl. Apr–May, fr. Aug–Oct.

# 枳 | *Poncirus trifoliata*

**别名：** 枸橘、臭橘　　　　　　　**科属：** 芸香科枳属

**形态特征：** 灌木或小乔木，高 1～6 m。小枝绿色，略扭扁，有枝刺，长 1～4 cm。三出复叶互生，近革质，叶总柄有翅，小叶无柄，倒卵形，长 1.5～6 cm，叶缘有波状浅齿，基部稍歪斜。两性花，白色，径 3.5～5 cm；雄蕊常 20 枚；雌蕊绿色，被毛。柑果球形，径 3.5～6 cm，果皮黄绿色，密生绒毛，芳香。花期 5—6 月，果期 10—11 月。

**习性、应用及产地分布：** 喜光，喜温和、湿润气候，耐寒、耐旱，不耐盐碱；发枝力强，耐修剪。果实外表似橘，为良好的观果树种，枝叶稠密兼具棘刺，常用作刺篱和屏障。产于中国河北、山东，南迄广东，黄河流域以南地区分布较为集中。

**Description:** Shrubs or trees to 1–6 m tall, deciduous. Branchlets green, with spines 1–4 cm. Leaves palmately 3-foliolate, subleathery, rachis narrowly winged, leaflet blade obovate, 1.5–6 cm long, margin crenate, base oblique. Flowers hermaphrodite, white, 3.5–5 cm in diam.; stamens usually 20, ovary green, hairy. Fruit globose, 3.5–6 cm in diam., pericarp yellow-green, densely tomentose, fragrant. Fl. May–Jun, fr. Oct–Nov.

# 柑橘 | *Citrus reticulata*

**科属：** 芸香科柑橘属

**形态特征：** 常绿小乔木，高 5～8 m。单身复叶，椭圆至椭圆状披针形，长 4～8 cm，宽 2～4 cm，翼叶线状或仅具痕迹，叶全缘或有细钝齿。花单生或 2～3 朵簇生于叶腋；花瓣 4～5 枚，白色，芳香；雄蕊 20～25 枚，花丝基部合生为束；花柱细长。果扁圆形，径 5～7 cm，橙红色或橙黄色，光滑，易剥离。花期 4～5 月，果期 10—12 月。

**习性、应用及产地分布：** 喜光，稍耐侧荫，喜温暖气候，不耐寒，忌积水。四季常青，春季白花芳香，秋季果实累累，为中国著名的果树之一，可植于庭园或风景区。中国为柑橘原产地，长江以南各省区广泛栽培。

**Description:** Small trees to 5–8 m tall, evergreen. Leaves 1-foliolate; leaf blade elliptic or ovate-lanceolate, 4–8 cm × 2–4 cm, basal articulated part to leaf blade usually linear or only a remnant, margin entire or apically obtusely crenulate. Flowers solitary to 2–3 in a fascicle, petals 4–5, white, fragrant, stamens 20–25, usually basally connate into bundles; style long, slender. Fruit oblate, 5–7 cm in diam., orange-red or orange-yellow, smooth, easily removed. Fl. Apr–May, fr. Oct–Dec.

# 酸橙 | *Citrus aurantium*

**科属：** 芸香科柑橘属

**形态特征：** 常绿小乔木，高6 m以上。枝三棱状，多棘刺。叶椭圆形或卵形，先端短尖，基部阔楔形，缘有波状钝齿；翼叶倒卵形，基部狭窄。总状花序兼有单花腋生；萼片4～5片，花后常增厚；花瓣白色，芳香，径2～3.5 cm；雄蕊20～25枚，基部合生成束。果圆球形至扁圆形，径7～8 cm，成熟时橙黄色，果肉味酸。花期5—6月，果期9—11月。

**习性、应用及产地分布：** 喜光，宜温暖、湿润气候，不耐寒，耐旱。著名的香花、观果树种，花色洁白如琼，果实瓣质深厚，香浓扑鼻，各地园林多作庭园或盆栽观赏。产于中国秦岭以南、长江下游地区，印度、日本、缅甸、越南亦有分布。

**Description:** Small evergreen trees, more than 6 m tall. Branches with spines. Leaf blade elliptic or ovate, apex shortly acuminate, base broadly cuneate, basal articulated part to leaf blade obovate, base narrow. Inflorescences racemes, with few flowers or flowers solitary; calyx lobes 4 or 5; petals white, fragrant, 2–3.5 cm in diam.; stamens 20–25, usually basally connate into bundles. Fruit globose to oblate, 7–8 cm in diam., orange-yellow when mature, sarcocarp acidic. Fl. May–Jun, fr. Sep–Nov.

# 甜橙 | *Citrus sinensis*

**别名**：橙子、广柑　　　　　　　**科属**：芸香科柑橘属

**形态特征**：常绿小乔木，高2～5 m。枝少刺或近无刺。单身复叶，卵状椭圆形至卵形，长6～10 cm，先端短尖，基部阔楔形，全缘；叶柄有狭翅。花单生或数朵簇生于叶腋，白色，长1.2～1.5 cm；雄蕊20～25枚；子房10～13室，花柱粗壮。果扁圆，径7～9 cm，熟时橙黄色。花期3—5月，果期10—12月。

**习性、应用及产地分布**：喜温暖环境，不耐寒，较耐荫，要求土质疏松、肥沃、透水透气性土壤。可种植于公园、庭院及亭、堂、院落角隅等处，或者草地边缘或湖、塘、池边。产于中国华东、华中、华南、西南至台湾等地。

**Description:** Small trees, evergreen, to 2–5 m tall. Branches with less spines or without spines. Leaves 1-foliolate, ovate-oblong to ovate, 6–10 cm, apex shortly acuminate, base broadly cuneate, margin entire, petiole with narrow wings. Flowers solitary or several clusters of leaf axils, white, 1.2–1.5 cm long; stamens 20–25; ovary 10–13 loculed, style stout. Fruit oblate, 7–9 cm in diam., orange-yellow when mature. Fl. Mar–May, fr. Oct–Dec.

# 34. 苦木科

## 臭椿 | *Ailanthus altissima*

**科属**：苦木科臭椿属

**形态特征**：落叶乔木，高可达30 m。树皮平滑而有直纹；枝条粗壮，叶痕大，倒卵形，内具9条维管束痕；无顶芽。奇数羽状复叶具小叶13～27对，连总柄在内长40～60 cm，总柄基部膨大；小叶近对生，卵状披针形，长7～13 cm，宽2.5～4 cm，先端渐尖，叶基部具1～2对腺齿，上面深绿色。顶生圆锥花序，长10～30 cm；花瓣5～6枚，淡绿色，长2～2.5 mm。翅果长椭圆形，扁平，种子位于中部，熟时黄褐色或红褐色。花期4—5月，果期8—10月。

**习性、应用及产地分布**：温带阳性树种。耐寒、耐盐能力强，不耐水湿，抗污染能力强；生长迅速。树干高大通直，冠大荫浓；春季嫩叶紫红，秋季红果满树，颇为美观，是良好的庭荫树、观赏树和行道树，有天堂树之美誉。中国黄河流域以南广泛分布。

**Description:** Trees, deciduous, up to 30 m tall. Bark smooth and straightly grained; branches stout, branchlets with big leaf scars, obovate, without terminal bud. Leaves odd-pinnate, leaflets 13–27 pairs, 40–60 cm long; leaflets ovate-lanceolate, apex shortly acuminate, base with 1–2 pairs of glands, adaxially dark-green. Panicles, terminal, 10–30 cm long; petals 5–6, pale green, 2–2.5 mm long. Samara flat, with seed in middle of wing, flat-globose, yellowish-brown or reddish-brown when mature. Fl. Apr–May, fr. Aug–Oct.

# 35. 楝科

## 苦楝 | *Melia azedarach*

**科属：** 楝科楝属

**形态特征：** 落叶乔木，高10 m。树皮灰褐色，浅纵裂，皮孔明显。2～3回奇数羽状复叶，互生，羽状复叶长20～40 cm；小叶卵形或卵状椭圆形，长3～7 cm，宽2～3.5 cm，先端渐尖，基部多少偏斜，叶缘有钝锯齿。二歧聚伞花序组成圆锥花序，腋生，花淡紫色或白色，长约1 cm，有香味；花瓣倒卵状匙形，外微被柔毛。核果球形至椭圆形，径1～1.5 cm，熟时黄色，宿存于树上，经冬不落。花期4—5月，果期10—12月。

**习性、应用及产地分布：** 热带及亚热带阳性树种。喜温暖、湿润气候，抗寒力弱，耐水湿，不耐干旱；生长迅速，寿命短，30～40年即衰老。树冠广卵形，羽状复叶较大，花淡雅芳香，适宜作庭荫树、行道树。产于河北南部、陕西南部、甘肃东南部及其以南各省低海拔地区，现引种栽培极为普遍。

**Description:** Trees, deciduous, up to 10 m. Bark brownish gray, longitudinally exfoliating, with obvious lenticels. Leaves odd-pinnate, 2- or 3-pinnate, 20–40 cm; leaflet blade ovate or ovate-elliptic, 3–7 cm × 2–3.5 cm, apex shortly acuminate, margin crenate. Panicles axillary, flowers purple or white, fragrant, petals obovate-spatulate. Drupe globose to ellipsoid, 1–1.5 cm in diam., yellow when mature, persistent. Fl. Apr–May, fr. Oct–Dec.

# 36. 漆树科

## 黄连木 | *Pistacia chinensis*

**别名：**楷木　　　　　　　　　　**科属：**漆树科黄连木属

**形态特征：**落叶乔木，高可达30 m。树冠近圆球形，树皮暗褐色，薄片状剥落，枝叶有特殊气味。偶数羽状复叶具小叶10～14枚，披针形或卵状披针形，长5～9 cm，宽1.5～2.5 cm，先端渐尖，全缘，基部偏斜。花单性，雌雄异株；先花后叶，雄花为紧密的总状花序，长6～7 cm，淡绿色；雌花为腋生的圆锥花序，长15～20 cm，紫红色。核果倒卵状球形，略扁，径约5 mm，初为黄白色，成熟时红色或蓝紫色。花期3—4月，果实9—11月成熟。

**习性、应用及产地分布：**温带及亚热带阳性树种。喜温暖，畏寒忌湿，对土壤要求不严；萌蘖力强，深根性，抗风；对二氧化硫、氯化氢和烟尘等抗性强。树冠浑圆，枝叶繁茂，早春嫩叶红色，入秋叶色深红或橙黄，优良色叶类树种。宜与槭类、枫香等混植于亭阁旁、草坪、坡谷、山石作庭荫树、行道树及大片风景树。产于中国长江以南各省区及华北、西北。

**Description:** Deciduous trees, up to 30 m tall. Crown subglobose; bark dark brown, flaking in pieces, emitting a special odor. Leaves even-pinnate, leaflet blade 10–14, lanceolate to ovate-lanceolate, 5–9 cm × 1.5–2.5 cm, apex shortly acuminate, margin entire, base oblique. Flowers unisexual, dioecious; female inflorescence panicles, axillary, 15–20 cm, purple-red. Drupe obovate-globose, white initial, red or blue-purple when mature. Fl. Mar–Apr, fr. Sep–Nov.

# 黄栌 | *Cotinus coggygria*

**别名：** 红叶树、烟树　　　　　　　　　**科属：** 漆树科黄栌属

**形态特征：** 落叶灌木或小乔木，高 3～6 m。树冠圆形；树皮暗灰褐色；小枝暗紫褐色，被蜡粉。叶倒宽卵形至宽椭圆形，长 3～8 cm，宽 2.5～6 cm，先端圆或微凹，基部圆形，无毛或仅叶背脉上被短柔毛，全缘；侧脉顶端常 2 叉状；叶柄长 1～4 cm，初霜来临时即变成红色。顶生圆锥花序，被柔毛，花瓣黄绿色；果序长 5～20 cm，多数不孕花淡紫色，羽毛状细长，花序梗宿存。核果肾形，径 3～4 mm，红色。花期 4—5 月，果熟期 7—8 月。

**习性、应用及产地分布：** 温带阳性树种。耐寒性强，耐干旱瘠薄，不耐水湿；生长快，根系发达，萌蘖性强；秋叶鲜红，与枫香、槭树等同为秋色叶类树种。园林中丛植于草坪、山坡或山石之侧或混植于其他红叶类或常绿树种中，层峦叠嶂、层林尽染，景色如画。产于中国陕西、山西、河北、河南、湖北、湖南等较干燥地区，常自成群落。

**Description:** Deciduous trees, 3–6 m tall. Crown globose, bark dark gray-brown, branchlets dark purple-brown. Leaf blades broadly elliptic to obovate, 3–8 cm × 2.5–6 cm, apex rounded to retuse, base rounded, glabrous or only pubescent abaxially, margin entire; petiole 1–4 cm long, turning red when frost. Panicles terminal, pubescent, petals yellow-green; infructescence 5–20 cm, most infecund, pale purple, persistent. Drupe reniform, 3–4 mm in diam., red. Fl. Apr–May, fr. Jul–Aug.

# 南酸枣 | *Choerospondias axillaris*

**别名：** 五眼果、酸枣　　　　**科属：** 漆树科南酸枣属

**形态特征：** 落叶乔木，高可达8～20 m。树皮灰褐色，条片状剥落；小枝褐色，具凸起皮孔。奇数羽状复叶常集生于小枝顶端，具小叶7～15枚，长25～40 cm；小叶对生，纸质，长圆形，长4～12 cm，宽2～4.5 cm，先端渐尖，基部偏斜，全缘。花单性或杂性异株；雄花序长4～10 cm；雌花单生于小枝上部叶腋，花瓣5枚，黄白色，覆瓦状排列；子房上位，5室，花柱5个，分离。核果卵状椭圆形，长2～3 cm；核骨质，顶端具5个萌发小孔，熟时黄色。花期4—5月，果期8—10月。

**习性、应用及产地分布：** 亚热带及热带阳性树种。喜湿润气候，不耐水涝；生长快，根系浅，不抗风。速生树种，树干通直，枝叶繁茂，秋叶金黄，适宜作行道树及庭荫树。产于中国长江以南及西南地区。

**Description:** Deciduous trees, 8–20 m tall. Bark gray-brown, peeling off in strips; branchlets brown, lenticellate. Leaves odd-pinnate, subterminal, leaflet blade 7–15, 25–40 cm long, leaflet blade opposite, papery, oblong-ovate, 4–12 cm × 2–4.5 cm, apex acuminate, base oblique, margin entire. Flowers unisexual or polygamous, male inflorescence 4–10 cm; female flowers solitary in axils of distal leaves, petals 5, yellow-white, superior ovary 5-locular, 5-style. Drupe ovate-ellipsoidal, 2–3 cm long. Fl. Apr–May, fr. Aug–Oct.

# 元宝枫 | *Acer truncatum*

**别名：** 元宝槭、平基槭、华北五角枫　　**科属：** 槭树科槭属

**形态特征：** 落叶小乔木，高可达 8～10 m。树皮灰褐色，浅纵裂。单叶对生，掌状 5 裂，深至中部，长 5～10 cm，宽 8～12 cm，中裂片有时 3 裂，叶基部常截形，全缘；主脉 5 条，掌状；叶柄长 3～5 cm。花杂性，顶生伞房花序；萼片 5 枚，黄绿色；花瓣 5 枚，黄白色。翅果扁平，成直角或钝角，翅与坚果近等长，成熟时淡黄色或带褐色。花期 4—5 月，果期 8—10 月。

**习性、应用及产地分布：** 弱阳性树种。喜侧方庇荫，喜温凉气候；较耐旱，不耐涝和瘠薄；萌蘖力强，深根性，抗风，对城市环境适应性强。树姿优美，冠大荫浓，叶形秀丽，秋季叶橙黄色或红色，是北方著名的秋色叶树种，适宜在堤岸、湖边、草坪及建筑物附近植作庭荫树、行道树或风景林伴生树种。主要分布于中国东北的南部地区、江苏的北部以及安徽的南部等地。

**Description:** Small trees, deciduous, 8–10 m tall. Bark grayish brown, slightly longitudinally fissured. Leaves opposite, palmate, 5-lobed, 5–10 cm × 8–12 cm, lobes triangular-ovate, base usually truncate, margin entire; petiole 3–5 cm. Flowers hermaphrodite; inflorescence terminal, corymbose; sepals 5, yellow-green; petals 5, yellow-white. Samaras compressed, with wings spreading at obtuse or right angles, pale yellow or with brown when mature. Fl. Apr–May, fr. Aug–Oct.

# 秀丽槭 | *Acer elegantulum*

**科属：** 槭树科槭属

**形态特征：** 落叶乔木，高9～15 m。树皮深褐色，嫩枝淡紫绿色。叶纸质，基部心形，宽大于长，通常5裂，边缘具细圆齿，无毛。花序圆锥状，花瓣5枚。翅果嫩时淡紫色，成熟后淡黄色，小坚果球形，径6 mm，翅张开近水平，连同小坚果长2 cm。花期5月，果期9月。

**习性、应用及产地分布：** 温带树种，弱度喜光，稍耐荫，喜温凉、湿润气候，对土壤要求不严；生长速度中等，深根性，抗风力强。秀丽槭的秋叶变亮黄色或红色，适宜做庭荫树、行道树及风景林树种。分布于中国东北、华北至长江流域。

**Description:** Deciduous trees, 9–15 m tall. Bark on trunk dark brown, new branchlets pale purple-green. Leaflet blade opposite, papery, base broadly cordate, with width greater than length, usually 5-lobed, margin serrulate, glabrous. Inflorescence paniculate, petals 5. Samara pale purple when young, pale yellow after mature, nutlets globose, 6 mm in diam., wings spreading obtusely, wing including nutlets 2 cm long. Fl. May, fr. Sep.

# 三角枫 | *Acer buergerianum*

**别名：** 三角槭、鸭脚枫　　　　　**科属：** 槭树科槭属

**形态特征：** 落叶乔木，高可达6～20 m。树皮灰黄色，薄长条状剥落；小枝纤细。叶互生，全缘，近革质，椭圆形或倒卵形，长4～10 cm，宽3～5 cm，先端3浅裂至叶长1/4，裂片前伸，近等大，基部近圆形或楔形，叶背具白粉或细毛；三出脉；叶柄长2.5～5 cm。顶生伞房花序，径约3 cm，花序梗长1.5～2 cm，被柔毛；萼片、花瓣各5枚。双翅果张开成锐角或近直角，翅果长2～2.5 cm，小坚果显著凸起。花期4—5月，果期9—10月。

**习性、应用及产地分布：** 喜光，喜温暖、湿润、通风良好的环境；较耐水湿，萌芽力强，耐修剪；根系发达，生长速度较快。干皮美丽，枝叶茂密，春季花色黄绿，夏季浓荫覆地，入秋叶色橙红，颇为美观，宜作庭荫树、行道树及湖岸、溪边护岸树栽植。产于中国长江流域中下游至广东、台湾各省。

**Description:** Deciduous trees, 6–20 m tall, deciduous. Bark gray-yellow, thin strips flaking; branchlets slender. Leaves alternate, margin entire, subleathery, elliptic or obovate, 4–10 cm × 3–5 cm, apex shallowly 3-lobed to 1/4 leaves, base subrounded or cuneate; leaf blade abaxially whitish or tomentose, primary veins 3, petiole 2.5–5 cm. Inflorescence terminal, corymbose, ca. 3 cm wide, pedicel 1.5–2 cm, pubescent, sepals 5, petals 5. Wings spreading at acute or right angles, wings 2–2.5 cm long, nutlets strongly convex. Fl. Apr–Mar, fr. Sep–Oct.

# 鸡爪槭 | *Acer palmatum*

**科属：** 槭树科槭属

**形态特征：** 落叶灌木或小乔木，高 5～6 m。树皮平滑，小枝灰紫色，纤细光滑。叶纸质，近圆形，径 7～10 cm，掌状 5～9 深裂，裂片披针形，先端锐尖，叶背仅脉腋被白色丛毛，边缘具尖锐细重锯齿；叶柄长 4～6 cm。顶生伞房圆锥花序，花序梗长 2～3 cm；萼片、花瓣各 5 枚，花紫色；小坚果球形，径 7 mm，显著凸起，翅果长 2～2.5 cm，张开成钝角，幼时紫红色，成熟时黄色。花期 5 月，果期 9 月。

**习性、应用及产地分布：** 温带弱阳性树种。耐半荫，耐寒性不强；生长速度中等偏慢。树姿优美，叶形秀丽，秋叶红色或古铜色，为优良的庭园观赏树种。产于中国华北、华东、华中地区，长江流域各地园林广泛栽植。

**Description:** Deciduous shrubs or small trees, 5–6 m tall. Bark smooth, branchlets grayish purple, slender and smooth. Leaf blade papery, suborbicular, 7–10 cm × 5–9 cm, 5–9 lobed, lobes lanceolate, apex abruptly cuspidate, margin irregularly doubly serrate; petiole 4–6 cm. Inflorescence corymbose-paniculate, terminal, pedicel 2–3 cm; sepals 5, petals 5, flowers purple; nutlets elliptic-convex, 7 mm in diam., samaras 2–2.5 cm long, wings spreading at obtuse angle, purple-red when young, yellow when mature. Fl. May, fr. Sep.

# 红枫 | *Acer palmatum* 'Atropurpureum'

**别名：** 紫红鸡爪槭　　　　　**科属：** 槭树科槭属

**形态特征：** 本种为鸡爪槭的栽培品种。形态特征与原种相似，主要区别为叶片终年红色或紫红色，枝条紫红色。

**Description:** This species is a cultivated variety of *Acer palmatum*. Morphological characteristics are similar to the original species, the main difference is red or purple leaves all year round, branchlets purplish red.

# 羽毛枫 | *Acer palmatum* 'Dissectum'

**别名**：细叶鸡爪槭　　　　**科属**：槭树科槭属

**形态特征**：落叶灌木。树冠开展；小枝略下垂，紫红色，老枝暗红色。单叶对生，叶裂片7～11枚，掌状深裂达基部，裂片狭长且有羽状细裂，叶缘具细尖齿。入秋逐渐转红。花紫色，翅果成钝角。花期5月，果期9月。

**习性与应用**：中性偏阴树种，喜温暖、湿润、凉爽气候环境。若成片种植，恰似霜叶红于二月花；若与众绿色植物配植，犹如万绿丛中一点红。

**Description:** Deciduous shrubs or small trees. Crown horizontal; branchlets slightly pendulous, purplish red, old branches dark red. Leaves subopposite, leaflets 7–11, palmately 7- or 11-lobed to base, lobes lanceolate, margin irregularly doubly serrate. Turn red in autumn. Flowers purple, wings spreading at obtuse angles. Fl. May, fr. Sep.

# 樟叶槭 | *Acer cinnamomifolium*

**科属：** 槭树科槭属

**形态特征：** 常绿乔木，高可达 10～20 m。树皮灰色，光滑，内皮层红棕色；幼枝淡紫褐色，有绒毛。单叶对生，革质，长椭圆形，长 7～12 cm，先端渐尖，全缘，表面绿色，叶背淡绿色，有白粉和绒毛；羽状脉，中脉及侧脉在表面下凹，在叶背凸起。圆锥花序顶生，有绒毛，果翅成直角或锐角。花期 3—4 月，果期 7—9 月。

**习性、应用及产地分布：** 性强健，喜温暖、湿润气候，耐半荫，不耐寒。树形优美，可作为景观树种。产于中国东南部至湖南、贵州等地。

**Description:** Trees to 10–20 m tall, evergreen. Bark gray, smooth; inner bark red-brown; branchlets pale purple-brown, tomentose. Leaves subopposite, leathery, oblong, 7–12 cm, apex acuminate, margin entire, adaxially green, abaxially light green, pruinose and tomentose; pinninerved, midvein and lateral veins concave adaxially, convex abaxially. Inflorescence corymbose-paniculate, terminal, tomentose, wings spreading at acute or right angles. Fl. Mar–Apr, fr. Jul–Sep.

# 茶条槭 | *Acer ginnala*

**科属：** 槭树科槭属

**形态特征：** 落叶小乔木，高6～9 m。小枝淡绿色，无毛，多年生枝黄褐色。单叶对生，卵状椭圆形，长6～10 cm，掌状3裂，中裂片较大，基部近圆形，叶缘有不规则重锯齿；叶柄及主脉常带紫红色。花序圆锥状，花杂性，黄绿色；果翅不开展，成锐角。花期5—6月，果期9月。

**习性、应用及产地分布：** 弱阳性。耐寒，耐干燥瘠薄，抗病力强；深根性，萌蘖性强。秋叶易变成红色，翅果成熟前红艳可爱，是良好的庭园观赏树。产于中国东北、内蒙古及华北，朝鲜、日本也有分布。

**Description:** Small trees, 6-9 m tall, deciduous. Branchlets palegreen, glabrous, perennial branches yellowish brown. Leaves opposite, ovate-oblong, 6-10 cm, 3-lobed, middle lobes big, base suborbicular, margin irregularly doubly serrate; petiole and main vein purple-red. Panicle inflorescence, flowers hermaphrodite, yellow-green; wings spreading at acute angle. Fl. May-Jun, fr. Sep.

# 苦茶槭 | *Acer ginnala* ssp. *theiferum*

**科属：** 槭树科槭属

**形态特征：** 本种是茶条槭的亚种，与原种形态特征相近。叶卵形至椭圆状卵形，叶片不裂或不明显3～5裂，叶缘有不规则重锯齿，叶背疏生白色柔毛。翅果较大，长2.5～3.5 cm，直立成锐角。花期5月，果期9月。

**习性、应用及产地分布：** 弱阳性，耐寒，耐干燥，忌水涝，抗烟尘。分布于中国湖北、湖南、广东、广西等地。

**Description:** A subspecies of *Acer ginnala*, with similar morphological characteristics to the original species. Leaf blade ovoid to ovate-oblong, slightly 3–5 lobed, margin irregularly doubly serrate, abaxially sparsely pubescent. Wings big, 2.5–3.5 cm long, spreading at right angle. Fl. May, fr. Sep.

# 羽叶槭 | *Acer negundo*

**别名：** 复叶槭、糖槭　　　　　　　　**科属：** 槭树科槭属

**形态特征：** 落叶大乔木，高20m。树冠圆球形；树皮暗灰色；小枝绿色，粗壮，常被白色蜡粉。奇数羽状复叶对生，具小叶3～7枚，小叶卵状椭圆形，长5～10cm，宽2～4cm，先端锐尖，基部广楔形或钝圆形，边缘具3～5个粗锯齿；叶脉在叶背显著，叶柄长5～7cm。花单性异株，先叶开放，黄绿色，被柔毛。小坚果长圆形，扁平，翅果呈锐角或近直角，果柄细长。花期4—5月，果期8—9月。

**常见栽培种：** '金叶'羽叶槭、'银边'羽叶槭、'金斑'羽叶槭、'花斑'羽叶槭等。

**习性、应用及产地分布：** 阳性树种。喜冷凉气候，稍耐水湿；适生于深厚肥沃、湿润土壤；萌蘖力强，生长较快，寿命较短；抗烟尘能力强。槭属中少数的复叶类树种。枝叶茂密，入秋叶色金黄，颇为美观，在北方多作庭院树或行道树，或与常绿树种配植。原产于北美，中国东北、华北、内蒙古、新疆等地有引种栽培，东北地区生长良好，在湿热的长江流域下游生长不良。

**Description:** Trees to 20 m tall, deciduous. Crown globose, bark dark gray, branchlets green. Leaves odd-pinnate, leaflets 3–7, leaflet blades ovate-oblong, 5–10 cm × 2–4 cm, apex abruptly cuspidate, base broadly cuneate or obtuse, margin 3–5 serrate. Flowers unisexual, yellow-green, pubescent. Nutlets oblong, compressed, wings spreading, peduncle slender. Fl. Apr–May, fr. Aug–Sep.

# 罗浮槭 | *Acer fabri*

**别名**：红翅槭　　　　　　　　**科属**：槭树科槭属

**形态特征**：半常绿乔木，高可达10 m。叶披针形至长椭圆披针形，长7～10 cm，全缘，先端渐尖，两面无毛，主脉在两面凸起；嫩叶淡红色。花萼紫色，花瓣白色，伞房花序。小坚果同翅长3～3.5 cm，翅果自幼至成熟均为紫红色。

**习性、应用及产地分布**：耐荫、耐寒。罗浮槭是新挖掘的优良绿化、美化树种，作第二层林冠配置最为理想，宜作风景林、生态林、四旁绿化树种。产于中国广东、广西、江西、湖北、湖南、四川，生于海拔500～1 800米的疏林中。

**Description:** Semi-evergreen trees, up to 10 m tall. Leaf blade lanceolate to oblong-lanceolate, 7–10 cm long, margin entire, apex acuminate, glabrous, main vein convex adaxially and abaxially; new leaves pale red. Sepals purple, petals white, inflorescence corymbose-paniculate. Small nuts with wings 3–3.5 cm long; wings purple-red.

## 37. 无患子科

### 复羽叶栾树 | *Koelreuteria bipinnata*

**别名：**黄山栾树　　　　　　　　　　**科属：**无患子科栾树属

**形态特征：**落叶乔木，高可达20 m以上。2回羽状复叶互生，羽叶长45～70 cm，具小叶9～15枚；小叶近纸质，卵状椭圆形，长3.5～8 cm，宽2～3.5 cm，先端渐尖，基部稍偏斜，叶背具短柔毛，边缘有细锯齿。圆锥花序顶生，长40～65 cm；花瓣4枚，黄色。蒴果卵状椭圆形，长4～7 cm，宽3.5～5 cm，具三棱，红色。花期7—9月，果期8—10月。

**习性、应用及产地分布：**温带及亚热带树种。喜光，耐半荫，耐寒；适生于石灰质土壤，耐干旱瘠薄；萌芽力强，深根性，生长速度中等。树形端正，枝叶茂密而秀丽，春季嫩叶多为红色，夏季黄花满树，入秋叶色金黄，是理想的庭荫树、行道树及园景树，也用作水土保持防护林、荒山绿化树种。产自中国东部、中南及西南部地区，是华北地区常见树种。

**Description:** Trees, deciduous, more than 20 m tall. Leaves 2-pinnate, alternate, 45–70 cm long, leaflets 9–15, blades papery, ovate-oblong, 3.5–8 cm × 2–3.5 cm, apex acuminate, base slightly oblique, abaxially densely pubescent, margin serrate. Inflorescence panicle, terminal, 40–65 cm long, petals 4, yellow. Capsules ellipsoid, 4–7 cm × 3.5–5 cm, 3-ridged, red. Fl. Jul–Sep, fr. Aug–Oct.

# 无患子 | *Sapindus mukorossi*

**科属：** 无患子科无患子属

**形态特征：** 落叶乔木，高可达20～25 m。树冠广卵形或扁球形；树皮灰白色，平滑不裂；枝开展，小枝无毛。偶数羽状复叶互生，具小叶8～14枚；纸质或薄革质，卵状披针形或卵状长椭圆形，长7～15 cm，宽2～5 cm，先端尖，基部不对称，全缘。圆锥花序顶生，长15～30 cm；花萼黄白色或淡紫色；花瓣5枚，披针形。核果近球形，径约2 cm，熟时黄色或橙黄色；种子球形，有光泽。花期5—6月，果期9—10月。

**习性、应用及产地分布：** 亚热带树种。喜光，稍耐荫，喜温暖、湿润气候，耐寒性差；生长较快，寿命长；深根性，抗风力强；萌芽力弱，不耐修剪。树形高大，绿荫稠密，秋叶金黄，颇为美观，适宜作庭荫树及行道树，若与其他秋色叶树种或常绿树种配植在草坪、路旁或建筑物附近，更可为秋景增色。产于中国长江流域及其以南各地。

**Description:** Trees, deciduous, up to 20–25 m tall. Crown broadly ovate or compressed-ellipsoid; bark grayish white, smooth; branches horizontal, branchlets glabrous. Leaves even-pinnate, alternate, leaflets 8–14; oblong-lanceolate or oblong-elliptic, 7–15 cm × 2–5 cm, apex acute, base asymmetrical, margin entire. Inflorescences terminal, conical, 15–30 cm long, sepals yellow-white or pale purple, petals 5, lanceolate. Drupe subglobose, 2 cm in diam., yellow or orange when mature; seeds globose, lustrous. Fl. May–Jun, fr. Sep–Oct.

# 38. 七叶树科

## 七叶树 | *Aesculus chinensis*

**科属：** 七叶树科七叶树属

**形态特征：** 落叶乔木，高25 m。树皮深褐色或灰褐色；小枝粗壮，黄褐色或灰褐色，具皮孔；冬芽大，有树脂。掌状复叶叶柄长10～12 cm，小叶通常7枚，纸质，倒卵状长椭圆形，长8～16 cm，宽3～5 cm，先端短锐尖，边缘具细锯齿；中脉在叶面显著，侧脉13～17对；小叶柄长5～17 mm。直立的大型圆锥花序顶生，近圆柱形，长21～25 cm，花小，白色；花瓣4枚，上方2枚具橘红色或黄色斑纹。蒴果球形，径3～4 cm，黄褐色，平滑；种子暗褐色，具大而明显的种脐痕。花期4—5月，果期9—10月。

**习性、应用及产地分布：** 亚热带阳性树种。喜温暖、湿润气候，稍耐寒。适生于深厚肥沃、排水良好的土壤。深根性，萌蘖力不强，不宜移植；生长偏慢，寿命长。树干通直，冠如华盖，叶形美丽，开花时硕大的花序直立于叶丛中，为世界著名的观赏树种，宜孤植或群植为庭荫树或行道树。中国河北南部、山西南部、河南北部、陕西南部均有栽培。

**Description:** Trees, deciduous, up to 25 m tall. Bark dark brown, lenticels. Digitate leaves, petiolules 10–12 cm long, leaf blade usually 7, papery, obovate-oblong, 8–16 cm × 3–5 cm, apex shortly acuminate, margin serrate; main vein obvious adaxially. Inflorescences terminal, conical, subterete, 21–25 cm long; flowers small, white. Capsule globose, 3–4 cm in diam., yellowish brown, smooth; seeds dark brown. Fl. Apr–May, fr. Sep–Oct.

# 39. 冬青科

## 枸骨 | *Ilex cornuta*

**别名**：鸟不宿　　　　　　　　　　**科属**：冬青科冬青属

**形态特征**：常绿灌木，高1～3 m。枝密集而开展，小枝粗，具纵脊。叶在枝上螺旋状排列，硬革质，长圆状四方形，长4～8 cm，宽2～4 cm，具尖硬刺齿5枚，叶端刺齿反曲，基部两侧各有1～2个大刺齿，基部圆形，上面深绿色且有光泽，边缘反卷。聚伞花序簇生于2年生小枝叶腋；花小，黄绿色，4基数。核果球形，径8～10 mm，熟时鲜红色。花期4—5月，果期10—12月。

**习性、应用及产地分布**：亚热带树种。喜光，也耐荫。适生于暖热、阴湿的气候条件，在酸性土中生长良好，不耐盐碱和瘠薄，较抗旱。生长缓慢，寿命长。叶形奇特，碧绿光亮；红果满枝，经冬不凋。宜孤植于花坛中心，对植于庭前、路口或丛植于草坪边缘、大乔木之下，也是很好的绿篱材料。产于中国江苏、上海、安徽、浙江、江西、湖北、湖南等地。

**Description:** Shrubs or small trees, evergreen, 1–3 m tall. Branches horizontal. Leaves radially spreading on branchlets, thickly leathery, quadrangular-oblong, 4–8 cm × 2–4 cm, obscure adaxially, margin with 1 or 2 spines per side, base rounded, leaf blade abaxially greenish, adaxially deep green, shiny, margin reflexed. Flowers small, yellow-green. Drupe globose, 8–10 mm in diam., red when mature. Fl. Apr–May, fr. Oct–Dec.

## 无刺枸骨 | *Ilex cornuta* var. *fortunei*

**科属：** 冬青科冬青属

**形态特征：** 常绿灌木，是枸骨的自然变种。叶厚革质，长圆形，全缘，偶有刺齿。可孤植、列植或与其他树种配植于公园、广场、庭院、道路及岩石园。

**Description:** Shrubs, evergreen, variant of *Ilex cornuta*. Leaf blade leathery, long-rounded, entire, rarely margin with spinose teeth. Planted alone, in rows or in combination with other tree species in parks, squares, courtyards, roads and rock gardens.

# 大叶冬青 | *Ilex latifolia*

**别名**：大苦丁、阔叶冬青    **科属**：冬青科冬青属

**形态特征**：常绿乔木，高可达10 m。树皮灰黑色，浅裂；小枝粗，无毛。叶厚纸质，长圆形或卵状长圆形，长8～19 cm，宽4.5～7.5 cm，先端钝或短渐尖，基部圆或宽楔形，缘有疏锯齿，上面深绿色，萌芽枝及新叶紫红色；中脉上面凹下，侧脉15～17对。由聚伞花序组成的假圆锥花序簇生于2年生枝叶腋，雄花序每个分支有花3～9朵，雌花序则具花1～3朵；花淡黄色，4基数。核果球形，径约7 mm，成熟时红色或棕红色，具褶皱和凹点，背面具1纵脊。花期4—5月，果期9—10月。

**习性、应用及产地分布**：耐荫，不耐寒。绿叶红果，颇为美观，宜作园林绿化及观赏树种。产于日本及中国长江中下游至华南地区。

**Description:** Trees, evergreen, 10 m tall. Bark gray-black, shallowly fissured; branches strong, glabrous. Leaf blade oblong or ovate-oblong, 8–19 cm × 4.5–7.5 cm, apex obtuse or shortly acuminate, base rounded or broadly cuneate, margin sparsely serrate, abaxially greenish, adaxially deep green; midvein impressed adaxially, lateral veins 15–17 pairs. Cymes or pseudopaniculate, axillary on second year's branchlets. Fruit red, globose, ca. 7 mm in diam. Fl. Apr–May, fr. Sep–Oct.

# 龟甲冬青 | *Ilex crenata* var. *convexa*

**别名：** 豆瓣冬青　　　　　　　**科属：** 冬青科冬青属

**形态特征：** 常绿灌木，高1～3 m，为钝齿冬青的栽培变种。叶厚革质，无毛有光泽，椭圆形至长倒卵形，长0.5～1.5 cm，先端圆钝，缘有浅钝齿，叶面凸起。花小，白色，成聚伞花序。果球形，径6～7 mm，熟时黑色。花期5—6月，果期8—10月。

**习性、应用及产地分布：** 庭院种植，可作盆景材料。主要分布于中国广东、福建、山东等地，日本也有分布。

**Description:** Shrubs, evergreen, 1–3 m tall. Leaf blade leathery, glabrous, lustrous, elliptic to obovate, 0.5–1.5 cm long, apex obtuse. Flowers, white, cyme. Fruit globose, 6–7 mm in diam., black when mature. Fl. May–Jun, fr. Aug–Oct.

# 40. 卫矛科

## 大花卫矛 | *Euonymus grandiflorus*

**科属：** 卫矛科卫矛属

**形态特征：** 半常绿小乔木，高可达10 m。树干灰褐色，小枝绿色；叶倒卵形至椭圆形，长4～10 cm，边缘具细尖锯齿，侧脉细密，革质。花黄绿色或黄白色，径可达2 cm；聚伞花序。蒴果近球形，黄色，具红色假种皮。花期5—6月，果期7—10月。

**习性、应用及产地分布：** 秋季叶常为紫红色，可作为园林绿化树种。产于中国西部或西南地区。

**Description:** Semi-evergreen small trees, up to 10 m tall. Twigs green. Leaf blade obovate to elliptic, 4–10 cm long, margin finely crenulate, leaf blade leathery. Flowers yellow-green or yellow-white, 2 cm in diam., cyme. Capsule globose, yellow, with red arillate. Fl. May–Jun, fr. Jul–Oct.

# 卫矛 | *Euonymus alatus*

**别名：** 鬼箭羽　　　　　　　　　　**科属：** 卫矛科卫矛属

**形态特征：** 落叶灌木，高1～3 m。小枝四棱形，具2～4条木栓质阔翅，宽可达1 cm；冬芽圆形。叶对生，倒卵形至椭圆形，长2～8 cm，宽1～3 cm，边缘具细尖锯齿；叶柄长1～3 mm。花通常3朵组成聚伞花序，花序梗长约1 cm；萼片半圆形；花瓣近圆形，淡绿色，径约8 mm。蒴果4深裂，有时仅1～3个心皮发育，棕紫色；种子椭圆形，长5～6 mm，褐色。花期5—6月，果期7—10月。

**习性、应用及产地分布：** 温带阳性树种。对气候和土壤适应性强，耐寒；耐干旱、瘠薄，在中性、酸性及石灰性土壤上均能生长；萌芽力强，耐修剪。枝翅奇特，早春嫩叶及秋叶均为紫红色，落叶后紫红色果实悬挂于枝间，颇为美观，园林中可作绿篱，孤植、丛植于草坪、斜坡、水边，或于山石间、林缘、亭廊边配置均适宜。除东北、新疆、青海、西藏、广东及海南以外，中国各省区均产。

**Description:** Deciduous shrubs, 1–3 m tall. Young branches, green, with 2 or 4 winglike corks, wings up to 1 cm wide. Leaves opposite, obovate or obovate-elliptic, 2–8 cm × 1–3 cm, margin crenulate to serrulate; petiole 1–3 mm long. Sepals suborbicular, flowers pale green, ca. 8 mm in diam. Capsule 4-lobed, reddish brown when fresh, only 1–3 lobes developed. Fl. May–Jun, fr. Jul–Oct.

## 丝棉木 | *Euonymus maackii*

**别名：** 白杜、华北卫矛　　**科属：** 卫矛科卫矛属

**形态特征：** 落叶小乔木，高可达 6 m。树皮灰褐色，老时纵沟裂；大枝开展，小枝细长，绿色，近四棱形，无木栓翅。叶菱状椭圆形至卵状椭圆形，长 4～8 cm，宽 2～5 cm，边缘具细锯齿。聚伞花序有花 3～7 朵；花部 4 基数，淡黄绿色，径约 8 mm；花药紫色。蒴果 4 深裂，径约 1 cm；种子长椭圆状，红色假种皮全部包被种子。花期 5—6 月，果期 9—10 月。

**习性、应用及产地分布：** 温带阳性树种。有一定耐寒力，耐水湿，较耐旱。根系发达，抗风；萌蘖力强，耐修剪。树冠饱满形美，枝叶娟秀细致，春季黄花满树，秋叶经霜转红，红色果实挂满枝梢。可作为庭院孤赏树、绿荫树或林缘、路旁、草坪、湖边、溪畔等处大片自然景观林、防风林及工厂、矿区绿化树种。产于中国中部、北部及东部各省，朝鲜、日本和俄罗斯亦有分布。

**Description:** Deciduous small trees, 6 m tall. Bark grayish brown, longitudinally fissured when old; branches green. Leaves, orbicular-ovate, 4–8 cm × 2–5 cm, margin crenulate. Flowers, yellow-green, cyme. Capsule with 4 deep grooves; seeds subglobose, dark brown, partially covered by orange aril. Fl. May–Jun, fr. Sep–Oct.

# 扶芳藤 | *Euonymus fortunei*

**别名**：蔓卫矛　　　　　　　　**科属**：卫矛科卫矛属

**形态特征**：常绿灌木至藤木植物，高一至数米。茎匍匐或攀缘；小枝绿色，近圆形，微有棱，密生小瘤状突起。叶薄革质，卵形或广椭圆形，长5.5～8 cm，宽1.5～4 cm，浓绿色，有时带紫色，边缘具钝齿。花小，绿白色，5～15朵或更多组成3～4歧聚伞花序；花部4基数。蒴果粉红色，近球形，径6～12 mm；种子长椭圆状，棕褐色。花期5—6月，果期10—11月。

**习性、应用及产地分布**：温带及亚热带阴性树种。喜温暖、阴湿环境，极耐荫，遮光70%仍可正常生长；耐寒性不强，对土壤要求不严。本种叶色油绿，入秋常转为红色，有极强的攀缘能力，常用以掩盖墙面、山石或老树干。产于中国黄河流域以南各省区，北京以南各城市均有栽培；朝鲜、日本亦有分布。

**Description:** Evergreen shrubs or vines, 1 m or taller. Ascending or procumbent on ground or rock; branches green, rounded. Leaves, leathery, ovate or broadly elliptic, 5.5–8 cm × 1.5–4 cm, deep green, sometimes purple, margin crenulate to serrate. Flowers small, green and white. Capsule pink, 6–12 mm in diam.; seeds brown to red-brown. Fl. May–Jun, fr. Oct–Nov.

# 大叶黄杨 | *Euonymus japonicus*

**别名**：冬青卫矛　　　　　　　　**科属**：卫矛科卫矛属

**形态特征**：常绿灌木或小乔木，高3～8 m。枝叶繁茂，小枝近四棱形。叶革质有光泽，倒卵形或狭椭圆形，长3～5 cm，宽2～3 cm，先端尖或钝，基部楔形，边缘有钝齿。聚伞花序腋生；花部4基数，绿白色，径5～7 mm；花盘肥大。蒴果近球形，径约8 mm，淡红色；种子椭圆状，长约6 mm，宽4 mm，棕色。花期3—4月，果期6—7月。

**常见变种、变型及栽培种**：'金边'大叶黄杨、'银边'大叶黄杨、'金心'大叶黄杨、'银心'大叶黄杨、'斑叶'大叶黄杨。

**习性、应用及产地分布**：温带及亚热带阳性树种。喜温暖、湿润气候，较耐寒，不耐水湿，极耐修剪。树形整齐，叶色光亮，四季常绿，与珊瑚树、'大叶'罗汉松同为"三大海岸绿篱"树种，可修整成各种几何体、动物或文字图案，环植于门旁、道边，列植于道路、亭廊两侧、建筑周围，或点缀于草地、台坡、桥头、树丛前和花坛中心，均甚美观。产于中国贵州、广西、湖南、江西等省市。

**Description:** Evergreen shrubs or small trees, 3–8 m tall. Young branches tetragonous. Leaf blade leathery, obovate or long elliptic, 3–5 cm × 2–3 cm, apex acuminate or obtuse, base cuneate, margin crenulate to serrate. Flowers, green or white, 5–7 mm in diam. Capsule subglobose, light red; seeds elliptic, 6 mm × 4 mm, brown. Fl. Mar–Apr, fr. Jun–Jul.

# 41. 黄杨科

## 小叶黄杨 | *Buxus microphylla*

**科属**：黄杨科黄杨属

**形态特征**：常绿灌木，高1～6 m。树皮黄褐色，浅纵裂；小枝绿色，四棱形。叶倒卵形至倒卵状椭圆形，长1～2.5 cm，通常中部以上最宽，先端微凹，基部楔形；侧脉两面不显；叶柄及叶下面中脉基部被毛。花黄绿色，密集头状花序腋生或顶生，簇生花序无花瓣，无花梗。蒴果近球形，径0.8～1 cm，熟时黄褐色或紫褐色；种子亮黑色。花期3—4月，果期10—11月。

**习性、应用及产地分布**：耐寒性较强，稍耐荫，是良好的绿篱与盆景树种。产于中国安徽黄山、浙江龙塘山、江西庐山和湖北神农架等地。

**Description:** Evergreen shrubs, 1–6 m tall. Bark shallow yellowish brown, branches green, tetragonous. Leaf blade ovate to obovate-elliptic, 1–2.5 cm long, emarginate, base cuneate; petiole and midrib below leaves tomentose. Flowers yellowish green, dense capitate axillary or terminal inflorescence, no peduncle. Capsule subglobose, 0.8–1 cm in diam., yellowish brown or purple-brown when ripe; seeds black. Fl. Mar–Apr, fr. Oct–Nov.

# 42. 鼠李科

## 枳椇 | *Hovenia acerba*

**别名：** 拐枣、万字果　　　　　　**科属：** 鼠李科枳椇属

**形态特征：** 落叶乔木，高可达 10～25 m。树皮灰褐色，不规则纵裂；幼枝红褐色，被棕褐色短柔毛。叶纸质，宽卵形或心形，长 8～17 cm，先端渐尖，基部截形或心形，下面沿脉或脉腋被毛，边缘具不整齐浅钝齿，近顶端锯齿不显；基出脉 3 条，不达齿端；叶柄具腺体。不对称二歧聚伞花序，结实时花序轴膨大肉质化，核果成熟时黄褐色，霜后味甜可食（俗称鸡爪梨）。花期 5—7 月，果期 8—10 月。

**习性、应用及产地分布：** 温带及亚热带阳性树种。喜温暖、湿润气候，有一定耐寒能力；对土壤要求不严，深根性，生长迅速，萌芽力强。树形端正，分枝均匀，叶大而光亮，果序轴肥厚扭曲，是结合园林生产的良好庭荫树、行道树。中国特有种，主要产于长江和黄河中下游地区。

**Description:** Deciduous trees, 10–25 m tall. Bark grayish brown, irregular longitudinal fissure; branchlets brown, with short brown pubescent. Leaf blade papery, broadly ovate or cordate, 8–17 cm long, apex acuminate, base truncate or cordate, margin finely serrulate, in upper or nearly terminal leaves conspicuously dentate, rarely subentire. Asymmetric inflorescences, dichasial cymose panicles, fruiting peduncles and pedicels dilated and fleshy. Fruit yellow-brown or brown at maturity. Fl. May–Jul, fr. Aug–Oct.

# 枣树 | *Ziziphus jujuba*

**别名：** 红枣　　　　　　　　　　　**科属：** 鼠李科枣属

**形态特征：** 落叶乔木，高可达10 m以上。树冠卵形；树皮灰褐色，条裂；枝红褐色，光滑，呈"之"字形弯曲，具2个托叶刺，长刺可达3 cm，粗直，短刺下弯，长4～6 mm。叶互生，卵形至椭圆状卵形，长3～7 cm，宽1.5～4 cm，先端渐尖，基部歪斜，有光泽，无毛，基出主脉3条，边缘具细锯齿。聚伞花序腋生；花小，两性，芳香；花盘厚。核果卵形至长圆形，长2～3.5 cm，熟时暗红色，中果皮肉质厚，味甜可食，果核坚硬。花期5—7月，果期8—9月。

**习性、应用及产地分布：** 温带强阳性树种。对环境适应性较强，喜干冷气候；耐干旱瘠薄，极不耐水湿；萌蘖力强，根系深广，抗风沙。枝干劲拔，翠叶垂荫，朱实累累，宜结合鲜果生产在庭园、路旁作庭荫树、园路树成片栽植。产于中国，除东北严寒地区和西藏外，全国其他各省均有种植，黄河中下游、华北平原栽培最为普遍。

**Description:** Deciduous trees, 10 m tall. Crown ovate; branches reddish-brown, flexuose, smooth, with 2 stipular spines, long spines erect, short spines recurved. Leaf blade papery, alternate, ovate or ovate-elliptic, 3–7 cm × 1.5–4 cm, apex acuminate, base oblique, lustrous, glabrous, 3-veined from base, margin crenate-serrate. Flowers small, bisexual. Fruit ovate to long ovate-elliptic, 2–3.5 cm long, dark red. Fl. May–Jul, fr. Aug–Sep.

## 马甲子 | *Paliurus ramosissimus*

**科属：**鼠李科马甲子属

**形态特征：**灌木或小乔木，高可达4～6 m。小枝灰色。叶阔卵形，长3～7 cm，宽2～5 cm，先端钝圆，基部宽楔形，稍偏斜；上面深绿色，有光泽，边缘有锯齿；叶柄被毛，基部具2个紫红色针状托叶刺。聚伞花序腋生，花萼杯状，5裂；花瓣淡黄绿色，短于萼片；雄蕊5枚，与花瓣近等长；花盘黄色，子房圆形，陷入花盘之内。核果杯状，被棕色或黄褐色绒毛；周围具3浅裂木栓质厚窄翅，果梗被棕褐色绒毛，果皮坚硬；种子扁圆形，紫红色或褐色。花期5—8月，果期9—10月。

**习性、应用及产地分布：**亚热带阳性树种。抗寒性强，耐干旱贫瘠；速生，病虫害少。木质坚硬、针刺密布，可作果园等场地围护绿篱。产于中国长江以南各省区海拔2 000 m以下山地，朝鲜、日本、越南亦有分布。

**Description:** Shrubs or small trees, 4–6 m tall. Leaf blade broadly ovate, apex obtuse, base broadly cuneate, symmetric to slightly oblique; upper blade dark green, lustrous, margin obtusely serrate or serrulate, petiole pubescent; stipular spines erect, dark red, 2 per node. Flowers in axillary cymes; petal yellowish green, shorter than sepals; stamens 5. Drupe cup-shaped, densely yellow-brown pubescent, margin distinctly 3-partite; seeds oblate, purple-red or red-brown. Fl. May–Aug, fr. Sep–Oct.

# 圆叶鼠李 | *Rhamnus globosa*

**科属：** 鼠李科鼠李属

**形态特征：** 落叶灌木，高可达2 m。小枝对生或近对生，灰褐色，有柔毛，顶端具针刺。叶近对生，倒卵形至椭圆形，先端突尖或钝；侧脉3～4对，在叶面下陷，叶背凸起，缘有细锯齿；幼叶被柔毛，后渐脱落。花通常簇生于叶腋，4基数。核果球形，径4～5 mm，成熟时黑色。花期4—5月，果期6—10月。

**习性、应用及产地分布：** 耐荫、耐干旱。可作为水土保持及林下绿化树种。产于中国内蒙古、华北至华东地区。

**Description:** Deciduous shrubs, 2 m tall. Branchlets opposite or subopposite, grayish brown, tomentose, apex spiny. Leaves opposite or subopposite, obovate or elliptic, apex abruptly cuspidate or obtuse; lateral veins 3 or 4 pairs, prominent abaxially, impressed adaxially, margin irregularly crenate-serrate; young leaves densely pubescent. Flowers in axillary. Fruit spherical, 4−5 mm in diam., black. Fl. Apr−May, fr. Jun−Oct.

# 长叶冻绿 | *Rhamnus crenata*

**科属：** 鼠李科鼠李属

**形态特征：** 灌木或小乔木，高4～7m。小枝褐色或紫红色，稍平滑，枝端刺状。叶纸质，近对生，在短枝上簇生，长圆形、椭圆形或倒卵状椭圆形，长4～15cm，先端突尖，基部楔形，边缘有圆齿状锯齿或细锯齿；侧脉7～12对。聚伞花序腋生，花瓣近圆形，顶端2裂；子房球形，花柱不裂。核果球形或倒卵状球形，绿色或红色，熟时黑或紫黑色，径6～7mm，具3分核，各有1粒种子。花期5—8月，果期8—10月。

**习性、应用及产地分布：** 亚热带阳性树种。喜光、耐荫，适应性强，较耐寒，耐干旱瘠薄；不择土壤；根系发达，枝叶繁茂，秋果密而多。园林中植于路边、林缘，也可作为坡地水土保持林树种，产于中国黄河流域以南。

**Description:** Shrubs or small trees, 4–7 m tall. Young branchlets brown or purple-red, slightly smooth, apex spiny. Leaf blade papery, obovate-elliptic, elliptic or obovate, 4–15 cm long, apex acuminate, base cuneate, margin finely crenate; lateral veins 7–12 pairs. Flowers in axillary cymes, petals subcircular, apex 2-lobed; ovary globose. Drupe globose or obovoid-globose, green or red, black, 6–7 mm in diam. Fl. May–Aug, fr. Aug–Oct.

## 43. 葡萄科

### 葡萄 | *Vitis vinifera*

**科属**：葡萄科葡萄属

**形态特征**：落叶木质藤本植物，茎长可达30 m。树皮红褐色，老时条状剥落；小枝光滑，卷须与叶对生。叶卵圆形，3～5掌状浅裂，长7～18 cm，宽6～16 cm，中裂片先端急尖，基部心形，两面无毛或下面稍被短柔毛，边缘具粗齿。圆锥花序与叶对生；花小，淡黄绿色。浆果球形或椭圆状球形，径1.5～2 cm，成串下垂，因品种不同，有白、青、红、褐、紫、黑等不同果色，被白粉；种子3～4枚，卵形或梨形。花期4—5月，果期8—9月。

**习性、应用及产地分布**：温带阳性树种。喜温暖、干燥、夏季高温的大陆性气候；耐干旱，怕涝；深根性，生长快。茎蔓柔韧，叶繁枝茂，可延蔓成荫，是优良的庭院棚架植物。原产于欧洲、西亚和北非一带，在中国各地均有栽培。

**Description:** Deciduous woody vines, 30 m long. Bark reddish-brown, strip exfoliation when old; young branchlets glabrous, tendrils bifurcate. Leaf blade oval, conspicuously 3−5 lobed or cleft, 7−18 cm × 6−16 cm, apex of midlobes acute, base deeply cordate, margin rough serrated teeth. Panicle opposite to leaves, flowers small, light yellowish green. Berry globose or elliptic, 1.5−2 cm in diam. Seeds 3−4, oval to pear-shaped. Fl. Apr−May, fr. Aug−Sep.

# 44. 杜英科

## 尖叶杜英 | *Elaeocarpus apiculatus*

**别名**：长芒杜英　　　　　　　　**科属**：杜英科杜英属

**形态特征**：常绿乔木，高可达10～30 m。树冠塔状圆锥形；树皮灰褐色，有皮孔。叶革质，有光泽，多聚生于枝端，绿叶丛中常存有少量鲜红老叶；叶片倒卵状长椭圆形，长15～25 cm，先端钝，中部以下渐窄，基部窄而钝，全缘或上半部具细锯齿；羽状网脉在下面凸起。有花5～14朵，总状花序多生于叶腋，花白色，径1～2 cm，悬垂，略香；花瓣先端呈撕裂状；花药顶端具长3～4 mm芒刺；子房3室。核果卵球形，长3～3.5 cm，熟时黄锈色。花期8—9月，果实冬季成熟。

**习性、应用及产地分布**：喜光，喜温暖、湿润环境；抗风，萌芽力强，生长快。树干挺拔，树冠壮观，是中国华南地区优良的园林风景树和行道树。

**Description:** Evergreen trees, 10–30 m tall. Tower-shaped conical crown; bark grayish brown, lenticels. Leaf blade leathery, lustrous, 15–25 cm long, apex obtuse, base narrow and obtuse, margin entire or serrulate, some red leaves often present on crown. Flowers white, 1–2 cm in diam., slightly sweet, anthers apex with spikes, 3–4 mm long, ovary 3-loculed. Drupe ovoid, 3–3.5 cm long. Fl. Aug–Sep.

## 中华杜英 | *Elaeocarpus chinensis*

**科属：** 杜英科杜英属

**形态特征：** 常绿小乔木，高可达7 m。幼枝有柔毛，后渐脱落。叶薄革质，卵状披针形，长5～8 cm，先端渐尖，基部圆；侧脉4～6对，网脉不显，有波状浅齿；叶柄长1.5～2 cm。总状花序生于无叶老枝上，花两性或单性；花瓣5枚，长圆形，先端不裂；雄蕊8～10枚，花丝极短。核果椭圆形，长不及1 cm。花期5—6月。

**习性、应用及产地分布：** 喜温暖、潮湿环境，耐寒性稍差。稍耐荫，根系发达，萌芽力强，耐修剪。分布于中国广东、广西、浙江、福建、江西、贵州、云南。生长于海拔350～850 m的常绿林中。

**Description:** Trees, evergreen, 7 m tall. Branchlets puberulent, glabrous. Leaf blade papery, ovate-lanceolate, 5–8 cm long, apex acuminate, base rounded; lateral veins 4–6 per side, veinlets inconspicuous on both surfaces, margin minutely crenate; petiole 1.5～2 cm long. Racemes in axils of fallen leaves, bisexual or unisexual; petals 5, oblong, margin nearly entire; stamens 8–10, filaments very short. Drupe ellipsoid, shorter than 1 cm. Fl. May–Jun.

# 45. 锦葵科

## 木芙蓉 | *Hibiscus mutabilis*

**别名：** 芙蓉花　　　　　　　　　　　　**科属：** 锦葵科木槿属

**形态特征：** 落叶灌木或小乔木，高2～6 m。小枝、叶、花梗、花萼、子房均密生星状绒毛。叶大，宽卵形或卵圆形，5～7掌状裂，长8～12 cm，宽10～15 cm，基部心形，边缘具钝圆齿；掌状脉，叶柄长5～20 cm。花单生或簇生于枝端，花梗长5～10 cm；花具副萼，小苞片8～10枚，线形，基部合生，密被星状绒毛；花萼钟形；花冠径约8 cm，初开时白色或淡红色，后变深红色，单瓣或重瓣。蒴果扁球形，径约2.5 cm。花期10—11月，果期12月。

**习性、应用及产地分布：** 亚热带树种。喜光，稍耐荫；喜温暖、湿润气候，不耐寒。花大而美丽，宜植于池岸，临水为佳；丛植于墙边、坡地、路边、林缘及建筑前。产于中国黄河流域及华东、华南各地，四川、云南、山东等地均有分布；日本和东南亚各国均有引种。

**Description:** Deciduous shrubs or small trees, 2–6 m tall. Branchlets, petioles, pedicel, epicalyx, and calyx densely stellate and woolly pubescent. Leaf blade broadly ovate to round-ovat, 5–7 lobed, 8–12 cm × 10–15 cm, base cordate, margin obtusely serrate; palmar vein, petiole 5–20 cm. Flowers solitary, axillary on upper branches; epicalyx lobes 8–10, calyx campanulate, corolla white, reddish or dark red, ca. 8 cm in diam. Capsule flattened globose, 2.5 cm in diam., yellowish hispid and woolly. Fl. Oct–Nov, fr. Dec.

# 木槿 | *Hibiscus syriacus*

**科属：** 锦葵科木槿属

**形态特征：** 落叶灌木，高 2～4 m。嫩枝密被黄色星状绒毛。叶菱形至三角状卵形，不裂或中部以上3裂，长3～6 cm，宽2～4 cm，先端钝，基部楔形，叶背沿脉疏生星状毛，边缘具不整齐圆锯齿；三出脉。花单生于叶腋，各部被星状短绒毛；副萼具线形小苞片6～8枚；花萼钟状，5裂；花冠钟形，径2～4 cm，淡紫、红、白等色，单瓣或重瓣，雄蕊不伸出花冠外。蒴果卵圆形，径约1.2 cm，密被黄色星状绒毛。花期7—10月，果期11月。

**习性、应用及产地分布：** 亚热带阳性树种。对环境的适应性很强，适生于温暖、湿润的气候，不耐干旱贫瘠；萌芽力强，耐修剪。花繁叶茂，花色丰富，园林中常植作花篱、花境等绿化材料，可丛植或单植以点缀庭园，也可于墙边、水滨种植。产于印度、叙利亚和中国中部；中国多数省区有野生或栽培品种，以华东及华南地区较为常见。

**Description:** Shrubs deciduous, 2–4 m tall. Branchlets yellow stellate puberulent. Leaf blade rhomboid to triangular-ovate or broadly lanceolate, variously 3-lobed or entire, 3–6 cm × 2–4 cm, apex obtuse to subacute, base cuneate, abaxially puberulent along veins, margin irregularly incised. Flowers solitary, stellate puberulent; epicalyx, lobes 6–8; calyx campanulate, 5-lobed; corolla campanulate, 2–4 cm in diam., blue-purple, violet, white or reddish. Capsule ovoid-globose, densely yellow stellate puberulent. Fl. Jul–Oct, fr. Nov.

# 海滨木槿 | *Hibiscus hamabo*

**科属：** 锦葵科木槿属

**形态特征：** 落叶丛生灌木，高1～4 m。树皮灰白色。叶厚纸质，近圆形，长3～6 cm，宽稍大于长，先端钝圆，具尖头，基部近圆形，两面密被灰白色星状毛，边缘有细锯齿。花单生于枝端叶腋；花萼5裂，副萼8～10裂；花冠钟状，径5～8 cm，花瓣金黄色，反卷，内面基部暗紫色。蒴果三角状卵形，密被黄褐色星状绒毛，径约1.5 cm，花萼宿存。花期6—10月，果期10—11月。

**习性、应用及产地分布：** 亚热带阳性树种。略耐干旱，耐短期水涝；对土壤适应能力强；根系发达，抗风。夏季开花，花色金黄，艳丽夺目，花期长而繁茂，秋季叶色橙黄，在阳光照耀下缤纷绚丽，是中国华东沿海地区优良的海岸防风林树种，也是优良的庭院观赏树种。产于中国浙江舟山群岛和福建沿海岛屿，日本、朝鲜亦有分布。

**Description:** Shrubs deciduous, 1–4 m tall. Bark grayish white. Leaves thickly papery, suborbicular, 3–6 cm long, slightly wider than long, apex obtuse, cuspidate, base suborbicular, densely grayish stellate pubescent, margin serrulate. Flowers solitary, axillary; calyx 5-lobed, epicalyx 8–10 lobed; corolla campanulate, 5–8 cm in diam., petals golden yellow, retrorse, inner base dark purple. Petals obovate, densely yellow-brown stellate tomentose, ca. 1.5 cm in diam., calyx persistent. Fl. Jun–Oct, fr. Oct–Nov.

## 46. 椴树科

### 心叶椴 | *Tilia cordata*

**别名**：欧洲小叶椴

**科属**：椴树科椴树属

**形态特征**：落叶乔木，高可达20～30 m。树冠圆球形，树干灰褐色；幼枝嫩时被柔毛，后脱落。叶卵圆形，长3～8 cm，先端渐尖，基部心形，稍偏斜，表面暗绿色，叶背苍绿色，脉腋处有棕色簇毛。花黄白色，芳香，无退化雄蕊，5～7朵呈聚伞花序；舌状苞片长4～8 cm，下部与花序轴合生。果实近球形，径约9 mm，表面被星状毛。花期6月，果期8—9月。

**习性、应用及产地分布**：喜光、耐寒，可忍耐-15℃低温；对烟尘和气体抗性强。树冠雄伟，夏日浓荫覆地，黄花满树，芳香浓郁，广泛用作行道树、庭荫树。原产于欧洲，中国新疆、南京、青岛等地有栽培。

**Description:** Trees, deciduous, 20–30 m tall. Crown globose, trunk grayish brown; young shoots pubescent when young, glabrescent. Leaves ovoid, 3–8 cm long, apex acuminate, base cordate, slightly oblique, adaxially dark green, abaxially pale green, vein axils with brown tufts. Flowers yellow and white, fragrant, without stamens, 5–7 in cymes; lingual bracts 4–8 cm long, lower part united with rachis of inflorescence. Fruit subglobose, ca. 9 mm in diam., surface stellate tomentose. Fl. Jun, fr. Aug–Sep.

# 47. 梧桐科

## 梧桐 | *Firmiana platanifolia*

**别名：** 青桐  **科属：** 梧桐科梧桐属

**形态特征：** 落叶乔木，高可达15 m。树皮青绿色，平滑。叶心形，径15～30 cm，3～5掌状裂，裂片三角形，先端渐尖，基部心形，两面无毛或略被短柔毛，全缘；基生脉7条；叶柄与叶片近等长。圆锥花序顶生，长20～50 cm；花淡黄绿色，萼片深裂几至基部。蓇葖果膜质，具柄，有毛；每个蓇葖果有种子2～4个，着生于果皮边缘，种子形如豌豆，径约7 mm，成熟时棕色，有皱褶。花期6—7月，果期9—11月。

**习性、应用及产地分布：** 温带及亚热带树种。喜光，喜温暖、湿润气候，耐寒性较强，较耐旱；萌芽力弱，生长迅速，浅根系。树冠有佳荫，干绿如翠玉，秋季叶片金黄，是优美的庭荫树。宜配植于庭院、草坪、池畔、湖边、坡地，也可作行道树。产于中国湖北西部及四川南部，华北、华南及西南地区栽培广泛，历史悠久；日本亦有分布。

**Description:** Trees, deciduous, 15 m tall. Bark greenish, smooth. Leaf blade cordate, 15−30 cm in diam., palmately 3−5 lobed, lobes triangular, apex acuminate, base cordate, both surfaces glabrous or minutely puberulent, entire, basal veins 7. Inflorescence paniculate, terminal, 20−50 cm long; calyx yellowish green, divided nearly to base. Follicle membranous, stalked, pilose, 2−4 seeded, bearing pericarp margins, seeds globose, ca. 7 mm in diam., brown at maturity, wrinkled. Fl. Jun−Jul, fr. Sep−Nov.

## 梭罗树 | *Reevesia pubescens*

**科属：**梧桐科梭罗树属

**形态特征：**常绿乔木，高可达15 m。幼枝被星状柔毛。单叶互生，卵状长椭圆形，长7～12 cm，先端渐尖，基部楔形，全缘，叶面疏生短柔毛，叶背密被黄褐色星状毛。花两性，花瓣5枚，白色，长1～1.5 cm，花丝合生成长管状，贴于花柱生长；聚伞花序顶生。蒴果木质，梨形，长2.5～3.5 cm，具5棱，密生淡褐色柔毛。花期5—6月，果期7—9月。

**习性、应用及产地分布：**树干直，枝繁叶茂，花序大而繁密，花朵醒目，可作行道树或庭荫树。产于中国海南、广西及西南地区，南京、上海地区已有栽培。

**Description:** Trees, evergreen, up to 15 m tall. Branchlets stellate puberulent when young. Leaves alternate, leaf blade ovate-oblong, 7–12 cm, apex acuminate, base obtuse, entire, adaxially sparsely puberulent, abaxially densely yellow-brown stellate tomentose. Flowers bisexual, petals 5, white, 1–1.5 cm, filaments connate into long tubular; inflorescence cymose, terminal. Capsule woody, pyriform, 2.5–3.5 cm, densely brownish puberulent. Fl. May–Jun, fr. Jul–Sep.

# 48. 瑞香科

## 结香 | *Edgeworthia chrysantha*

**别名：** 黄瑞香、打结花、三桠　　　　**科属：** 瑞香科结香属

**形态特征：** 落叶灌木，高1～2 m。小枝褐色，粗壮而柔软，可打结，三叉分枝，具皮孔，叶痕大。叶常集生于枝顶，椭圆状倒披针形，长6～20 cm，宽2～7 cm，先端急尖，基部楔形，叶背被细长柔毛，脉上毛较密；侧脉9～11对；叶柄短。花30～50朵成头状花序，花序梗密被柔毛；花黄色，芳香，无柄；花萼筒状，盛开时长约2 cm，外面密被白色丝状毛，内面黄色。核果椭圆形，径约3.5 mm，绿色。花期2—3月，果期7—8月。

**习性、应用及产地分布：** 亚热带树种。喜半荫，耐寒性较差。肉质根，不耐水湿，忌盐碱土。株型优雅，早春开花，花多成簇，芳香浓郁，适合孤植、列植或丛植于庭园、水边、道旁、墙隅，或点缀于假山、岩石之间。中国南部大部分省区均有野生或栽培品种，日本及美国东南部亦有栽培。

**Description:** Shrubs, deciduous, 1–2 m tall. Branchlets brown, strong and soft, knotted, trifurcate branched, lenticellate, leaf scars large. Leaves clustered at the apex of the branches, elliptic oblanceolate, 6–20 cm × 2–7 cm, apex apiculate, base cuneate, leaf back covered with slender pilose; lateral veins 9–11 pairs, petiole short. 30–50 flowered, capitates, yellow, fragrant, sessile. Drupe ellipsoid, ca. 3.5 mm in diam., green. Fl. Feb–Mar, fr. Jul–Aug.

# 49. 胡颓子科

## 胡颓子 | *Elaeagnus pungens*

**科属：** 胡颓子科胡颓子属

**形态特征：** 常绿灌木，高可达4 m，具刺。小枝褐色，密被锈色鳞片。叶革质，椭圆形，长5～10 cm，宽1.8～5 cm，先端钝，基部楔形或圆形，边缘反卷或波状。叶面有光泽，幼时具银白色和褐色鳞片，后脱落，叶背银白色，密被盾状褐色鳞片。花白色，1～3朵生于小枝叶腋，芳香，下垂；萼筒漏斗状圆筒形，密被黄褐色鳞片。果椭圆形，长1.2～1.4 cm，幼时被褐色鳞片，熟时红色。花期9—12月，果期翌年4—6月。

**习性、应用及产地分布：** 喜光，耐半荫；喜温暖气候，不耐寒。耐干旱贫瘠，也耐水湿。对土壤要求不严。枝叶扶疏，叶下面银色，花香果红，适于庭园、公园中群植或草坪、花丛中丛植，也可作绿篱或点缀于池畔、石间。产于中国长江流域以南各省，日本亦有分布。

**Description:** Shrubs, evergreen, 4 m tall, spines frequent. Young branches brown, densely ferruginous scaly. Leaves leathery, elliptic, 5–10 cm × 1.8–5 cm, apex obtuse to bluntly acute, base rounded, margin reflexed or undulate. Leaves, leathery, adaxially glabrous and glossy, abaxially with dense whitish and scutellate brown scales. Flowers white, 1–3 in axils of branchlets, fragrant, pendulous; calyx tube funnelform, cylindrical, densely yellow-brown scales. Drupe oblong, 1.2–1.5 cm. Fl. Sep-Dec, fr. Apr-Jun of following year.

# 金边胡颓子 | *Elaeagnus pungens* 'Goldrim'

**科属：** 胡颓子科胡颓子属

**形态特征：** 常绿灌木，此种为胡颓子的栽培品种，叶缘亮黄色，颇具光泽。

**Description:** Evergreen shrubs, a cultivated variety of *Elaeagnus pungens*, having a glossy yellow leaf margin.

## 木半夏 | *Elaeagnus multiflora*

**科属：** 胡颓子科胡颓子属

**形态特征：** 落叶灌木，高 2～3 m。枝红褐色，常无刺。叶椭圆状卵形，长 3～7 cm；幼叶表面有星状柔毛，后渐脱落，叶背银白色且有褐斑。花常单生于叶腋，白色，有香气，花冠筒与裂片近等长。果椭球形，长 1.2～1.8 cm，红色，果梗细长。花期 4—5 月，果期 6 月。

**习性、应用及产地分布：** 性强健，喜光，喜温暖、肥沃土壤。果红艳美丽，果期长，宜植于园林绿地观赏。产于中国长江中下游地区，日本也有分布。

**Description:** Shrubs, deciduous, 2–3 m tall. Branches reddish brown, often spineless. Leaves elliptical ovate, 3–7 cm long; leaves stellate puberulent when young, glabrescent with age, leaf back silvery white with brown spots. Flowers often solitary in axil, white and fragrant, corolla tube subequal to lobes. Drupe ellipsoidal, 1.2–1.8 cm, red, fruiting pedicel slender. Fl. Apr–May, fr. Jun.

# 50. 柽柳科

## 柽柳 | *Tamarix chinensis*

**别名：** 观音柳　　　　　　　　**科属：** 柽柳科柽柳属

**形态特征：** 落叶灌木或小乔木，高可达3～8 m。树皮暗褐红色，光亮；小枝暗紫红色，稠密纤细，常开展而下垂。叶细小，鳞片状，长1～3 mm，密生。总状花序组成圆锥花序生于顶端；花瓣5枚，粉红色，直生或略开展。蒴果3裂，长3 mm。花期4—9月，有时春、夏、秋开3次花；果期10月。

**习性、应用及产地分布：** 亚热带及温带阳性树种。耐烈日暴晒，耐寒；抗旱能力极强，耐水湿；极耐盐碱。根系发达，萌芽力强，耐修剪。枝条细柔，姿态婆娑，绿叶与粉红花相映成趣，花期长，多栽植于庭院，或配植于其他落叶乔木下，亦适宜温带海滨、河畔等处盐碱地。分布几乎遍及中国各地，常见于河流冲积平原、海滨、滩头、潮湿盐碱地和沙荒地。

**Description:** Deciduous shrubs or small trees, 3–8 m tall. Bark dark reddish brown, shiny; branchlets dark red-purple, dense and slender, often spreading and pendulous. Leaves fine, scaly, 1–3 mm, dense. Racemes forming panicles at apex; petals 5, pink, straight or slightly spreading. Capsule 3-lobed, 3 mm long. Fl. Apr–Sep, sometimes spring, summer, autumn bloom 3 times; fr. Oct.

# 51. 千屈菜科

## 紫薇 | *Lagerstroemia indica*

**别名：**痒痒树、百日红、满堂红　　　　**科属：**千屈菜科紫薇属

**形态特征：**落叶灌木或小乔木，高可达3～7 m。树皮灰色或灰褐色，薄片状脱落后光滑；枝干多扭曲，小枝四棱形，纤细无毛。叶对生或近对生，纸质，椭圆形，长2.5～7 cm，宽1.5～4 cm，先端短尖或钝，有时微凹，基部宽楔形或近圆形，无毛或下面脉上被毛，全缘，近无柄。花淡红色、紫色或白色，径3～5 cm；圆锥花序顶生，长7～20 cm；萼筒平滑无纵棱，三角形，直立；花瓣6枚，长1.2～2 cm，边缘皱缩。蒴果近球形，6瓣裂。花期6—9月，果期9—12月。

**习性、应用及产地分布：**喜光；喜温暖、湿润气候，耐寒性、耐旱性不强，耐盐碱；萌蘖性强，幼年生长迅速，寿命可达数百年。树姿优美，古朴典雅，树干光洁，"人若搔之，则枝干无风而自动，俗呼怕痒树"。孤植或丛植于草坪、林缘，配植于水溪、池畔；矮生品种可作绿化隔离带或花篱。产于亚洲，广植于热带各地；中国华东、华中、华南、西南各省区均有栽培。

**Description:** Deciduous shrubs or small trees, 3–7 m tall. Bark gray or grayish-brown, flaky smooth after shedding; branches twisted, slender, glabrous and 4-angled. Leaves opposite or subopposite, elliptic, 2.5–7 cm × 1.5–4 cm, apex acute or obtuse, sometimes concave, base broadly cuneate or rounded, entire, subsessile. Flowers reddish, purple or white, 3–5 cm in diam., terminal panicles, 7–20 cm long, petals 6, 1.2–2 cm, margin corrugated. Capsule subglobose, 6-lobed. Fl. Jun–Sep, fr. Sep–Dec.

## 福建紫薇 | *Lagerstroemia limii*

**别名**：浙江紫薇　　　　　　　　　**科属**：千屈菜科紫薇属

**形态特征**：落叶小乔木，树皮脱落而成褐色，光滑。叶互生至近对生，近革质，顶端短渐尖，叶面光滑，叶背脉腋处被柔毛。圆锥花序顶生，花轴及花梗密被柔毛；花瓣淡红色至紫色，圆卵形，有皱纹；蒴果卵形，长8～12 mm，成熟时褐色，光亮，约1/4包藏于宿存萼内。花期5—6月，果期7—8月。

**习性、应用及产地分布**：喜温暖，稍耐荫，有一定抗寒能力，在北京可露地栽培。常孤植或丛植于草坪、林缘处。产于中国福建、浙江、湖北等地。

**Description:** Small trees, deciduous. Bark peeling off and turning brown, smooth. Leaves alternate or nearly opposite, subleathery, apex shortly acuminate, glossy, abaxially puberulous on veins axils. Panicle terminal, rachis and pedicels densely puberulous; petals reddish to purple, ovate, crinkled; capsule ovate, 8–12 mm long, brown at maturity, shiny, ca. 1/4 enclosed in persistent calyx. Fl. May–Jun, fr. Jul–Aug.

# 52. 桃金娘科

## 大叶桉 | *Eucalyptus robusta*

**别名**：桉树、红桉

**科属**：桃金娘科桉属

**形态特征**：大乔木，高可达25～30 m。树皮暗褐色，厚而松软，纵裂粗糙不脱落；小枝淡红色。幼叶厚革质，卵形，长达11 cm；成熟叶为革质，卵状长椭圆形至广披针形，长8～18 cm，先端长尖，基部圆形，两面具油腺点；侧脉多而细，具边脉。4～12朵花成腋生伞形花序，花冠圆锥形，径1.5～2 cm。蒴果卵状壶形，径0.8～1 cm。花期4—5月和8—9月，花后约3个月果熟。

**习性、应用及产地分布**：热带树种。喜光；喜温暖、湿润气候，不耐低温，耐水湿，不耐干瘠。根系深，萌芽力强，生长迅速。树姿秀丽，树干粗直，叶疏而下垂，为热带及亚热带地区首选庭荫树、行道树，亦可作为河畔、湖旁及沿海地区低湿处防护林树种。原产于澳洲沿海地区沼泽地，中国浙江、四川及华南地区各省均有引种。

**Description:** Trees up to 25–30 m tall. Bark dark brown, thick and soft, longitudinally fissured, rough not shed; branchlets pale red. Young leaves thickly leathery, ovate, 11 cm long; mature leaves ovate-oblong to broadly lanceolate, 8–18 cm long, apex long acute, base rounded, both surfaces glandular, lateral veins numerous and thin. Inflorescences axillary, umbels 4–12 flowered, corolla conical, 1.5–2 cm in diam. Capsule ovate-shaped, 0.8–1 cm in diam. Fl. Apr–May and Aug–Sep.

# 红千层 | *Callistemon rigidus*

**别名**：红瓶刷、金宝树　　　　　　　　**科属**：桃金娘科红千层属

**形态特征**：常绿灌木，株高2～3 m。树皮灰褐色，嫩枝具棱。叶硬革质，条形，长3～8 cm，宽2～5 mm，叶面光滑，有透明腺点，全缘；中脉显著，无叶柄。穗状花序似瓶刷状，生于枝端，长约10 cm；花瓣绿色，有油腺点；雄蕊多数，鲜红色，花药暗紫色；花柱先端绿色，较雄蕊稍长。蒴果半球形，先端平，径7 mm。花期6—8月，果期秋至冬季。

**习性、应用及产地分布**：喜光，喜暖热气候，耐45℃高温和-10℃低温，在华南、西南地区可露地过冬；对水分要求不严，耐干旱瘠薄；萌发力强，耐修剪；抗大气污染。株形飒爽美观，每年春末夏初火树红花，满枝吐焰，盛开时千百枚雄蕊组成一支支艳红的瓶刷子，甚为奇特；花期长，易于栽培，为庭园树、绿化隔离带首选，亦可于城镇近郊荒山或森林公园等处植为风景树、防风林。原产于澳洲，中国华南地区有栽培。

**Description:** Shrubs, evergreen, 2–3 m tall. Bark gray-brown, young branchlets ribbed. Leaves hard leathery, strip-shaped, 3–8 cm × 2–5 mm, smooth, transparent glandular points, entire; midrib conspicuous, without petiole. Spikes like bottle brushes, born on branches, ca. 10 cm long; petals green, with oil glands; stamens numerous, bright red, anthers dark purple; style apex green, slightly longer than stamens. Capsule hemispherical, apex flat, 7 mm in diam. Fl. Jun–Aug, fr. from autumn to winter.

# 轮叶赤楠 | *Syzygium buxifolium* var. *verticillatum*

**别名：** 三叶赤楠　　　　　　　　　　　**科属：** 桃金娘科蒲桃属

**形态特征：** 灌木或小乔木，高1.5～2 m。嫩枝红褐色，四棱形。3叶轮生，倒卵状椭圆形，长1.5～3 cm，宽1～2 cm，先端圆钝，基部阔楔形。花小，白色，聚伞花序顶生；萼筒倒圆锥形，长约2 mm；花瓣4枚，白色，分离，与萼筒等长；雄蕊长5 mm，与花柱等长。果球形，径4～5 mm。花期6—8月，果期9—10月。

**习性、应用及产地分布：** 喜光，较耐荫，喜温暖、湿润气候；生长缓慢，侧根发达，萌蘖力强，耐修剪。生长缓慢，适合植作绿篱、园景树。产于中国浙江、江西、湖南、广东、广西等地。

**Description:** Shrubs or small trees, 1.5−2 m tall. Young branchlets reddish-brown, 4-angled. Leaves whorls on shoots, obovate-elliptic, 1.5−3 cm × 1−2 cm, apex obtuse, base broadly cuneate. Flowers small, white, cymes terminal, calyx tube inverted conical, ca. 2 mm long. Petals 4, white, separate, as long as calyx tube. Fruit spherical, 4−5 mm in diam. Fl. Jun−Aug, fr. Sep−Oct.

# 香桃木 | *Myrtus communis*

**别名：** 茂树、香叶树　　　　**科属：** 桃金娘科香桃木属

**形态特征：** 常绿灌木，高1～3 m。小枝密集，灰褐色，枝四棱，嫩枝有锈色毛。叶对生或3叶轮生，革质，揉搓后具香味，卵形至披针形，长2.5～5 cm，宽0.5～1 cm，先端渐尖，基部楔形，叶面深绿色，有光泽；冬季幼叶紫红色，中脉被柔毛。花常单生于叶腋，白色或淡红色，径1.5～2 cm，芳香；花瓣5枚，倒卵形，被腺毛；雄蕊多达50枚，离生，与花瓣等长。浆果圆形或椭圆形，径6～8 mm，蓝黑色，顶部具宿萼。花期5—6月，果期11—12月。

**习性、应用及产地分布：** 热带阳性树种。喜温暖、湿润气候，萌芽力强，耐修剪；病虫害少。花纯白色，形似梅花，盛花时枝条缀满花朵，芳香宜人，清雅而脱俗，有"爱神木""银香梅"和"祝福木"之美誉。适于作居住区、道路、花境、林缘和向阳围墙前的背景树、高篱，形成绿色屏障。原产于地中海地区，中国南部有栽培。

**Description:** Shrubs, evergreen, 1–3 m tall. Branchlets dense, gray-brown. Leaves opposite or whorls on shoots, leathery, fragrant, ovate to lanceolate, 2.5–5 cm × 0.5–1 cm, apex acuminate, base cuneate, leaves dark green, shiny; leaves turning purplish red in winter, midrib densely pilose. Flowers often solitary in axil, white or pale red, 1.5–2 cm in diam., aromatic. Petals 5, obovate, glandular hairs; stamens up to 50, free. Fruit round or elliptic, 6–8 mm in diam., blue-black, with a sepal at the top. Fl. May–Jun, fr. Nov–Dec.

## 花叶香桃木 | *Myrtus communis* 'Variegata'

**科属：** 桃金娘科香桃木属

**形态特征：** 本种为香桃木的栽培品种，形态特征和生态习性与原种相近。主要形态差异是叶片为斑叶。

**Description:** This species is a cultivated variety of *Myrtus communis*. The morphological characteristics and ecological habits are similar to the original species. The main difference is that the leaves are variegated.

## 小叶香桃木 | *Myrtus communis* var. *microphylla*

**科属：** 桃金娘科香桃木属

**形态特征：** 本种为香桃木的变种；与原种的区别是，枝繁叶茂，叶小，线形至披针形，长 1～2.5 cm，宽2～5 mm。

**Description:** This species is a variety of *Myrtus communis*. The main difference is that the leaves are luxuriant, small, linear to lanceolate, 1–2.5 cm × 2–5 mm.

## 菲油果 | *Feijoa sellowiana*

**别名：** 肥吉果、非油果　　　　　　　　**科属：** 桃金娘科南美棯属

**形态特征：** 常绿灌木或小乔木，高达5 m。叶对生，椭圆形或椭圆状倒卵形，长达4～8 cm，全缘，叶面绿色有光泽，背面密生白色绒毛。花单生于叶腋，径4 cm；花萼4裂；花瓣4枚，肉质，被白色绒毛，内面紫红色；雄蕊多数，细长花丝与花柱皆暗红色。浆果卵状椭球形，绿色稍到红色，被白粉，长5～7.5 cm，有宿萼。

**习性、应用及产地分布：** 喜光，喜温暖、湿润气候。树姿优美，花色艳丽，果可食；可做园林绿化及观赏树种。原产于南美洲亚热带国家，中国上海与云南地区有引种。

**Description:** Evergreen shrubs or small trees, 5 m tall. Leaves opposite, elliptic or elliptic obovate, 4–8 cm long, entire, adaxially green and glossy, abaxially covered with dense white pubescent. Flowers solitary in axil, 4 cm in diam.; calyx 4, split; petals 4, fleshy, densely white pubescent, inner purplish red; stamens numerous, slender filaments and styles are dark red. Berry ovate-ellipsoidal, slightly green to red, with white farinose, 5–7.5 cm, with persistent calyx.

# 松红梅 | *Leptospermum scoparium*

**别名：** 澳洲梅、鱼柳梅    **科属：** 桃金娘科松红梅属

**形态特征：** 常绿灌木，高可达2 m。枝纤细，红褐色，有柔毛。单叶互生，线形至披针形，长达1.2 cm，全缘，嫩叶有柔毛。花单生于叶腋，径达1.2 cm；花瓣5枚，白色或粉红色；雄蕊多数；蒴果木质，5瓣。花期晚秋至春末。

**习性、应用及产地分布：** 喜光，喜凉爽湿润气候；夏季怕高温和烈日暴晒，耐旱性较强，对土壤要求不严。品种丰富，花有单瓣、重瓣，花色有白、紫、红、粉等色，极富观赏价值，可用作庭院观赏灌木。原产于新西兰与澳大利亚。

**Description:** Shrubs, evergreen, 2 m tall. Branches slender, reddish brown, pilose. Leaves alternate, linear to lanceolate, 1.2 cm long, entire, young leaves pilose. Flowers solitary in axil, 1.2 cm in diam.; petals 5, white or pink; stamens numerous; capsule woody, 5-valved. Flowering from late autumn to spring.

# 53. 石榴科

## 石榴 | *Punica granatum*

**别名：** 安石榴、海榴  **科属：** 石榴科石榴属

**形态特征：** 落叶灌木或小乔木，高3～6 m。树干灰褐色，具瘤状突起；嫩枝具棱，枝常有刺。叶纸质，在长枝上对生，在短枝上簇生，长披针形至长卵形，长2～8 cm，宽1～2 cm，先端短尖或微凹，基部楔形至钝形，上面具光泽。花大，朱红色，1～5朵着生在1年生枝顶端及叶腋；萼筒钟形，萼片5～7裂，红色或淡黄色，质厚。浆果近球形，径5～12 cm，外种皮古铜色或古铜红色，可食，具宿存花萼。花期5—6月，果期9—10月。

**习性、应用及产地分布：** 热带、亚热带及温带阳性树种。喜温暖、湿润气候，较耐寒，有一定耐旱力；生长速度中等。花开于夏季，绿叶荫荫之中灿若红霞，入秋花实并丽，是美丽的观赏树种。宜孤植或丛植于庭院、游园之角，对植于门庭出口处，列植于溪边、坡地和建筑物之旁，绚烂之极。原产于伊朗、阿富汗等国家。

**Description:** Deciduous, shrubs or small trees, 3–6 m tall, glabrous. Branches sometimes spiny. Petiole 2–10 mm; leaf blade adaxially shiny, lanceolate, elliptic-oblanceolate, or oblong, 2–8 cm × 1–2 cm, base attenuate, apex obtuse or mucronate. Floral tube red-orange or pale yellow, campanulate-urceolate; sepals 5–7, erect, bright red-orange. Fruit globose, leathery berries, 5–12 cm in diam., crowned by persistent sepals. Seeds obpyramidal within juicy sarcotestal layer, ruby-red, pink, or yellowish white. Fl. Mar–Jun, fr. Sep–Oct.

# 54. 蓝果树科

## 喜树 | *Camptotheca acuminata*

**别名：**旱莲　　　　　　　　　　　　　　　**科属：**蓝果树科喜树属

**形态特征：**落叶乔木，高可达20 m以上。当年生小枝黄绿色，髓心片状分隔。叶纸质，卵状椭圆形，长8～20 cm，宽6～12 cm，先端渐尖，基部宽楔形，叶背疏被柔毛，脉上尤密，全缘；羽状脉在叶面下凹明显；叶柄长1.5～3 cm，带红晕。花杂性同株；头状花序球形，具长总梗。果矩圆形香蕉状，长2～2.5 cm，常多数集生成球状，熟时黄褐色。花期5—7月，果期9—11月。

**习性、应用及产地分布：**亚热带速生树种。喜光，喜温暖、湿润气候，不耐寒，不耐干旱贫瘠，较耐水湿；深根性，萌芽性强。树形高大雄伟，树姿端直，树冠宽广，枝叶繁茂，果形奇特，为南方广为栽培的优良庭荫树、行道树及平原湖区农田防护林树种。中国特有种，产于中国河南、长江流域以南各省区。国家二级重点保护植物。

**Description:** Trees, deciduous, more than 20 m tall. Bark light gray, deeply furrowed. Leaf blade abaxially greenish and lucid, oblong-ovate, oblong-elliptic, or orbicular, 8–20 cm × 6–12 cm, papery, slightly pubescent, base subrounded, margin entire, apex acute; petiole length 1.5–3 cm, with blush. Flowers heterozygous, capitate subglobose, long stalk. Fruit thinly winged, gray-brown, smooth and lucid when dry. Fl. May–Jul, fr. Sep–Nov.

# 55. 山茱萸科

## 洒金东瀛珊瑚 | *Aucuba japonica* var. *variegata*

**别名：** 花叶青木　　　　　　　　　　**科属：** 山茱萸科桃叶珊瑚属

**形态特征：** 常绿灌木，高达5 m。小枝绿色，无毛。叶对生，革质，卵状椭圆形，长6～14 cm，宽3～8 cm，先端渐尖，基部阔楔形，叶面有黄色斑点，两面有光泽，边缘中上部具2～6对疏粗齿，基部全缘。花小，紫红色或暗紫色；圆锥花序密生刚毛。果卵圆形，长1.2～1.5 cm，径5～7 mm，鲜红色。花期3—4月。

**习性、应用及产地分布：** 耐荫，喜温暖、湿润气候；生长势强，耐修剪；病虫害极少，对烟害抗性强。枝叶繁茂，四季常青，是珍贵的耐荫观叶树种，南方常植作绿篱，置于林缘树下或丛植于庭院一角，效果甚佳。产于中国浙江南部及台湾；朝鲜、日本亦有分布，世界各地广为栽培。

**Description:** Evergreen shrubs, 5 m tall. Branchlets green, glabrous. Leaves opposite, leathery, oval to elliptical, leaf blade 6–14 cm × 3–8 cm, variegated with yellow spots, abaxially light green, adaxially shiny green, apex pointed and blunt, base subrounded or broadly cuneate, margin with 2–6 pairs of teeth on upper half. Flowers small, purplish red or dark purple; panicles densely bristly. Fruit ovoid, 1.2–1.5 cm long, 5–7 mm in diam., bright red. Fl. Mar–Apr.

# 红瑞木 | *Cornus alba*

**别名**：红梗木、红瑞山茱萸　　　　**科属**：山茱萸科梾木属

**形态特征**：落叶灌木，高可达3 m。树皮紫红色；小枝近四棱形，血红色，常被白粉。叶对生，卵形至椭圆形，长5～8.5 cm，宽1.8～5.5 cm，先端突尖，基部近圆形，叶面暗绿色，叶背粉绿色，两面疏被柔毛，全缘；弧形侧脉4～6对。花白色或淡黄白色，伞房聚伞花序，径3 cm。花萼、花瓣及子房疏被短柔毛。核果长圆形，微扁，径5.5～6 mm，成熟时白色稍带蓝紫色，花柱宿存，核两侧压扁状。花期6—7月，果期8—10月。

**习性、应用及产地分布**：喜光，耐半荫、耐水湿、耐寒。干枝秋季红色，颇为美观，植于草坪、林缘及河岸、湖畔均甚适宜。产于中国东北、华北及西北地区。

**Description:** Deciduous shrubs, 3 m tall. Bark purple; branchlets almost 4-angled, blood-red. Leaves opposite, entire, pinkish green, ovate to elliptic, 5–8.5 cm × 1.8～5.5 cm, base suborbicular, sparsely pilose on both sides; curved lateral veins, 4–6 pairs. Flowers white or yellowish, cymose. Calyx, petals and ovary sparsely pubescent. Drupe oblong, blue-purple, slightly flat, 5.5–6 mm in diam. Fl. Jun–Jul, fr. Aug–Oct.

# 光皮梾木 | *Cornus wilsoniana*

**别名：** 光皮树　　　　　　　　　　　　　**科属：** 山茱萸科梾木属

**形态特征：** 落叶乔木，高可达8～15 m。树皮绿白至青灰色，薄片状剥落，光滑。叶对生，椭圆形，长6～12 cm，宽2～5.5 cm，先端渐尖，基部楔形，叶面暗绿色，叶背淡绿色，密被乳头状小突起及灰白色短柔毛；侧脉3～4对。花白色，径7 mm，顶生聚伞花序，径6～10 cm。果实圆球形，径6～7 mm，熟时紫黑色；核球形。花期5—6月中旬，果期10—11月。

**习性、应用及产地分布：** 较喜光，耐寒；深根性，萌芽力强，生长较快，寿命较长。树冠舒展，树皮斑驳，枝叶茂盛，夏季银花满树，是较理想的庭园树，亦作行道树。产于中国秦岭至淮河流域以南地区，集中于长江流域至西南各省。

**Description:** Deciduous trees, 8–15 m tall. Bark greenish-white to bluish-gray, flaky and smooth. Leaves opposite, elliptic, 6–12 cm × 2–5.5 cm, apex acuminate, base cuneate, leaf surface dark green, densely papillately pubescent and grayish pubescent; lateral veins 3–4 pairs. Flowers white, 7 mm in diam., terminal cymes, 6–10 cm in diam. Fruit spherical, 6–7 mm in diam., purple-black. Fl. May–Jun, fr. Oct–Nov.

## 香港四照花 | *Dendrobenthamia hongkongensis*

**科属：** 山茱萸科四照花属

**形态特征：** 常绿乔木，高可达6～20 m。树皮深灰色或黑褐色，平滑；幼枝和幼叶疏被褐色细毛。叶对生，革质，椭圆形至长椭圆形，长6～13 cm，宽3～6 cm，先端短渐尖或短尾状，基部宽楔形或钝尖形；中脉明显，侧脉3～4对。花50～70朵成球形头状花序，径1 cm，花瓣淡黄色，有香味，外有花瓣状白色大形苞片4枚。聚合果球形，径2.5 cm，被白色细毛，成熟时红色。花期5—6月，果期11—12月。

**习性、应用及产地分布：** 初夏白色总苞片覆盖满树，光彩夺目，是一种美丽的园林观赏树种。产于中国华东、华南、中南及西南各省区。

**Description:** Evergreen trees, 6–20 m tall. Bark dark gray or dark brown, smooth; young branches green or purplish green, sparsely pubescent with brown appressed trichomes. Leaves opposite, leathery, elliptic to oblong, 6–13 cm × 3–6 cm, apex shortly acuminate or shortly culate, base broadly cuneate or obovately lanceolate; middle vein distinct, lateral vein 3–4 pairs. Flowers 50–70 forming capitulum, petals yellowish, scented, with petal-like white large bracts. Aggregated fruit, spherical-shaped, 2.5 cm in diam., turning red when mature. Fl. May–Jun, fr. Nov–Dec.

## 山茱萸 | *Macrocarpium officinale*

**科属：** 山茱萸科山茱萸属

**形态特征：** 落叶灌木或小乔木，高可达4～10 m。老枝灰褐色，嫩枝绿色。叶对生，纸质，卵状椭圆形，长5.5～10 cm，宽2.5～4.5 cm，先端渐尖，基部近圆形，全缘，弧形侧脉6～7对，叶背脉腋处被黄褐色柔毛。花小，黄色，先叶开放，成伞形花序生于小枝顶端，花序梗粗壮，微被灰色短柔毛；花瓣卵形，微反卷，长3～3.5 mm；花盘肉质。核果椭圆形，长约2 cm，红色至紫红色。花期3—4月，果期9—10月。

**习性、应用及产地分布：** 喜光，稍耐荫；喜温暖气候，耐寒性强，较耐湿。春季黄花满树，入秋果实殷红，鲜艳可人，经冬不凋；适于庭院、亭际、园路转角、公园或自然风景区丛植。产于长江流域及河南、陕西等地。

**Description:** Deciduous shrubs or small trees, 4–10 m tall. Old branches gray-brown. Leaves opposite, papery, ovate-elliptic, 5.5–10 cm × 2.5–4.5 cm, apex acuminate, base suborbicular, entire, curved lateral veins 6–7 pairs, axils of lateral veins with dense light brown long soft trichomes. Flowers small, yellow, umbels, peduncle thick, slightly gray pubescent, petals ovate. Drupe oval, ca. 2 cm long, red to purplish red. Fl. Mar–Apr, fr. Sep–Oct.

# 56. 五加科

## 八角金盘 | *Fatsia japonica*

**别名：** 八角盘、手树　　　　　　　　　　**科属：** 五加科八角金盘属

**形态特征：** 常绿灌木，高可达5 m。幼枝、叶和花序均具易脱落的褐色毛。叶革质，近圆形，径20～40 cm，叶面有光泽，掌状5～9裂，裂片卵状长椭圆形，有锯齿，基部心形或截形；叶柄长10～30 cm。球状伞形花序组成圆锥花序，花序梗长20～40 cm；花小，花瓣卵状三角形，长3～4 mm，黄白色。浆果球形，径约5 mm，成熟时紫黑色。花期10—11月，果期翌年4—5月。

**习性、应用及产地分布：** 亚热带阴性树种。植于阳光直射处，则叶片萎缩；喜温暖、湿润气候，耐寒性不强，不耐干旱；抗污染。良好的观叶树种。在日本有"庭树下木之王"的美誉，最适植于庭前、窗下、门旁、墙隅及建筑物阴面。原产于日本，中国长江流域以南各地常见栽培。

**Description:** Evergreen shrubs, up to 5 m tall. Young shoots, leaves and inflorescences have brown hairs that fall off easily. Leaves leathery, suborbicular, 20−40 cm in diam., leafy glabrous, palmately 5-9-lobed, ovate to long ellipsose, serrate. Cymose umbels composed of panicles; flowers small, petal ovate-triangulate, yellow-white. Berry globose, ca. 5 mm in diam., purple-black when mature. Fl. Oct−Nov, fr. Apr−May of following year.

## 熊掌木 | *Fatshedera lizei*

**别名：** 常春金盘　　　　　　　　　　**科属：** 五加科熊掌木属

**形态特征：** 常绿灌木，高可达2 m。茎幼时具锈色柔毛，后渐脱落。单叶互生，掌状3～5裂，裂达近中部，叶长达20 cm，革质，深绿色，有光泽。花黄绿色，径约1 cm，由多数伞形花序组成顶生圆锥花序，不结果。

**习性、应用及产地分布：** 性耐荫，喜冷凉、湿润气候，不择土壤，抗污染和盐风。本种是八角金盘与大西洋常春藤的杂交种，英国育成，中国有引种。

**Description:** Evergreen shrubs, up to 2 m tall. Young stem, rust-colored pilose, then gradually falls off. Leaves alternate, palmate 3–5-lobed, leathery, dark green, shiny. Flower, yellow-green, umbels, 1 cm in diam.

# 57. 八角枫科

## 八角枫 | *Alangium chinense*

**科属：** 八角枫科八角枫属

**形态特征：** 落叶乔木或灌木，高可达3～5 m。幼枝紫绿色，具稀疏柔毛。单叶互生，纸质，近椭圆形、卵形，长13～19 cm，宽9～15 cm，顶端短锐尖，基部两侧不对称，叶面深绿色，无毛，叶背淡绿色，近无毛；掌状脉3～5条；叶柄长2.5～3.5 cm，紫绿色或淡黄色。聚伞花序腋生，长3～4 cm，有花7～30朵，花冠圆筒形，长1～1.5 cm，花瓣6～8枚，线形，长1～1.5 cm，反卷，基部合生，初为白色，后变为黄色；雄蕊和花瓣同数。核果卵圆形，径5～8 mm，幼时绿色，成熟后黑色。花期5—7月，果期7—11月。

**习性、应用及产地分布：** 阳性树种，稍耐荫。可作庭荫树。产于中国黄河、长江流域至华南、西南地区。

**Description:** Deciduous trees or shrubs, 3–5 m tall. Young branches purple-green, sparsely pilose. Leaves alternate, nearly elliptical, ovate, 13–19 cm × 9–15 cm, base asymmetrical, dark green, glabrous; palmate veins 3–5; petiole length 2.5–3.5 cm. Cyme, 7–30 flowers, corolla cylindrical, 6–8 petals, initially white, then yellow. Drupe, oval, 5–8 mm in diam., green when young and turning black when ripe. Fl. May–Jul, fr. Jul–Nov.

# 58. 大风子科

## 山桐子 | *Idesia polycarpa*

**科属：** 大风子科山桐子属

**形态特征：** 落叶乔木，高可达15 m。干皮灰白色，不裂。单叶互生，广卵形，长10～20 cm，先端渐尖，基部心形，基出掌状脉5～7条，缘有疏齿，叶面深绿色，叶背白色；叶柄长6～12 cm，基部有2～4个紫色、扁平腺体。花单性异株或杂性，无花瓣，萼片5枚，黄绿色，芳香，成顶生圆锥花序。浆果球形，红色，径7～8 mm。花期5—6月，果期9—10月。

**习性、应用及产地分布：** 树冠端整，秋日红果累累，甚为美观。宜作庭荫树或观赏树。产于中国华东、华中、西北及西南各地。

**Description:** Deciduous trees, up to 15 m tall. Bark gray-white, not cracked. Leaves alternate, broadly ovate, 10–20 cm long, apex acuminate, base heart-shaped, palmate vein 5–7, margin sparse, leaf surface dark green. Petiole, 6–12 cm, with 2–4 purple and flat glands. Flowers unisexual or heterozygous, without petals, sepals 5, yellow-green, aromatic. Berry, spherical, red, 7–8 mm in diam. Fl. May–Jun, fr. Sep–Oct.

# 59. 清风藤科

## 泡花树 | *Meliosma cuneifolia*

**科属：** 清风藤科泡花树属

**形态特征：** 落叶灌木或乔木，高可达9 m。树皮灰褐色；小枝暗褐色，无毛。单叶互生，纸质，倒卵形或倒卵状椭圆形，长8～12 cm，宽2.5～4 cm，先端短渐尖，中部以下渐狭，叶柄长1～2 cm。圆锥花序顶生，直立，长15～20 cm，被短柔毛；花瓣5枚，近圆形，长1～1.2 mm，雄蕊伸出花冠以外，长1.5～1.8 mm。核果扁球形，径6～7 mm。花期6—7月，果期9—11月。产于中国华中、西北及西南地区。

**Description:** Deciduous shrubs or trees, 9 m tall. Bark gray-brown; branchlets dark brown, glabrous. Leaves alternate, papery, obovate or obovate-elliptic, 8–12 cm × 2.5–4 cm, apex short and acuminate, petiole 1–2 cm. Panicles terminal, erect, 15–20 cm long, pubescent; petals 5, suborbicular. Drupe, oblate, 6–7 mm in diam. Fl. Jun–Jul, fr. Sep–Nov.

## 红枝柴 | *Meliosma oldhamii*

**别名：** 羽叶泡花树　　　　　　　**科属：** 清风藤科泡花树属

**形态特征：** 落叶乔木，干皮灰褐色。奇数羽状复叶，小叶7～15枚，小叶薄纸质，基部小叶卵形，中部小叶长圆状卵形至窄卵形，小叶先端尖或渐锐尖，基部楔形，叶缘疏生锐齿。圆锥花序直立，花白色，外面3片花瓣近圆形，内面2片花瓣稍短于花丝；子房被黄色柔毛。核果球形，径4～5 mm；核具网纹，中肋隆起。花期5—6月，果期8—9月。

**习性、应用及产地分布：** 亚热带树种。喜光也耐荫，喜温暖、湿润气候环境及深厚肥沃的湿润土壤，不耐寒。树干端直，冠枝横展，花序宽大，花白果红，是良好的园林观赏和绿荫树种。分布于中国长江流域各省和台湾。

**Description:** Deciduous trees, bark gray-brown. Odd-pinnately compound leaves, leaflets 7–15, thinly sessate. Panicles erect, flowers white, petals 3, nearly round, ovate pubescent. Drupe, spherical, 4–5 mm diam. Fl. May-Jun, fr. Aug-Sep.

# 60. 唇形科

## 水果蓝 | *Teucrium fruticans*

**科属：** 唇形科香科科属

**形态特征：** 常绿灌木。叶对生，全缘，卵圆形，长 1～2 cm，宽 1 cm。小枝四棱形，全株被白色绒毛，以叶背和小枝最多。花唇形，蓝色。花期6—7月。

**习性、应用及产地分布：** 喜光，耐干旱瘠薄。萌蘖力很强，耐修剪，可用作规则式园林的矮绿篱，为庭院带来一抹靓丽的蓝色。原产于地中海地区及西班牙。

**Description:** Evergreen shrub. Leaves opposite, entire, ovoid, 1-2 cm × 1 cm. Branchlets quadrangular, the whole plant is densely covered with white tomentose. Flower, labiates, blue. Fl. Jun-Jul.

# 61. 杜鹃花科

## 杜鹃花 | *Rhododendron simsii*

**别名：** 映山红、照山红　　　　　　　**科属：** 杜鹃花科杜鹃花属

**形态特征：** 落叶或半常绿灌木，高 1～3 m。小枝及叶密被黄褐色糙伏毛。叶常集生于枝顶，卵状椭圆形或椭圆状披针形，长 2～6 cm，先端短渐尖，叶面深绿色，叶背密被褐色糙伏毛，边缘微反卷。花 2～3（6）朵簇生于枝端，花冠宽漏斗形，径约 4 cm，蔷薇色或深红色，上方裂片具深红色斑点；雄蕊 10 枚，与花冠近等长；花柱伸出花冠外。蒴果卵形，长约 1.2 cm，密被糙伏毛。花期 4—6 月，果期 10 月。

**习性、应用及产地分布：** 喜光，稍耐荫，较耐寒，适生于温润、排水良好的酸性土壤。春日红花鲜艳夺目，可用于布置园林或点缀风景区。原产于日本，中国长江以南各省多栽培。

**Description:** Deciduous or semi-evergreen shrubs, 1–3 m tall. Branchlets and leaves densely covered with densely shiny brown appressed-setose. Leaves, ovate-elliptic or elliptic-lanceolate, 2–6 cm long, apex short acuminate, leaf surface dark green, margin slightly revolute. Flowers, 2–3 (or 6) clustered on top of branches, corolla broadly funnel-shaped, ca. 4 cm in diam., rose or crimson, upper lobes with deep red spots, stamens 10, styles extending out of corolla. Capsule ovate, ca. 1.2 cm long, densely obscured. Fl. Apr–Jun, fr. Oct.

## 毛白杜鹃 | *Rhododendron mucronatum*

**别名：** 白花杜鹃、琉球杜鹃　　　　**科属：** 杜鹃花科杜鹃花属

**形态特征：** 半常绿灌木，高1～2 m。多分枝，枝叶及花序梗均密生粗毛，芽鳞外被黏质。叶纸质，长椭圆形，长3～5 cm，宽1～2 cm，先端钝尖至圆形，基部楔形，网脉在叶背凸起。花白色，稀淡红色，芳香，1～3朵簇生于枝端；花萼大，绿色；花冠阔漏斗状，长3～4.5 cm，径约5 cm；雄蕊10枚，与花冠近等长；子房5室，密被毛，花柱远伸出花冠之外。蒴果长卵形，长约1 cm。花期4—5月，果期6—7月。

**习性、应用及产地分布：** 喜光，稍耐荫，适生于温润、排水良好的土壤。中国各大城市均有栽培，日本、越南、英国、美国广泛引种栽培。

**Description:** Semi-evergreen shrubs, 1–2 m tall. More branches, branches and peduncles, pubescence. Leaves, papery, oblong, 3–5 cm × 1–2 cm, apex blunt to round, base cuneate, veins bulging on the back of blade. Flowers white, faint red, aromatic, 1–3 clustered on top of branches, calyx large, green, corolla broadly funnelform, ca. 5 cm in diam., stamens 10, styles extending beyond the corolla. Capsule, ovate, ca. 1 cm long. Fl. Apr–May, fr. Jun–Jul.

## 羊踯躅 | *Rhododendron molle*

**别名：** 黄花杜鹃、闹羊花　　　　**科属：** 杜鹃花科杜鹃花属

**形态特征：** 落叶灌木，高 0.5～2 m。分枝稀疏，幼枝密被灰白色柔毛及疏刚毛。叶纸质，长圆形至长圆状披针形，长 5～11 cm，宽 1.5～3.5 cm，先端钝，具短尖头，基部楔形，边缘具睫毛，近全缘；中脉及侧脉在下面凸起，中脉被黄褐色刚毛；顶生伞形总状花序，先花后叶或花与叶同放，花及蒴果被微柔毛及疏刚毛；花冠宽钟形，金黄色或橙黄色，长 4.5 cm，径 6 cm；雄蕊 5 枚，不等长，不超过花冠；花柱长 6 cm，伸出花冠。蒴果圆柱形，长 2.5～3.5 cm。花期 3—5 月上旬，果期 7—10 月。

**习性、应用及产地分布：** 亚热带阳性树种。不耐寒，对土壤要求不严。花大美丽，常在园林中作为配景点缀；"叶有毒，羊食其叶，踯躅而死，故名羊踯躅"。分布于中国长江流域至广东各省。

**Description:** Deciduous shrubs, 0.5–2 m tall. Branches are sparse, young branches densely covered with sparsely bristles. Leaf papery, oblong to oblong-lanceolate, 5–11 cm × 1.5–3.5 cm, apex obtuse, shortly pointed, base cuneate, margin ciliate, suborbate. Flowers, umbels, corolla golden or orange-yellow, 4.5 cm long, 6 cm in diam.; stamens 5, not exceedingly corolla. Capsule cylindrical, 2.5–3.5 cm. Fl. Mar–May, fr. Jul–Oct.

# 马醉木 | *Pieris japonica*

**别名：** 日本马醉木、梫木　　　　　　　**科属：** 杜鹃花科马醉木属

**形态特征：** 常绿灌木，高约4 m。树皮棕褐色；小枝开展，无毛。叶簇生于枝顶，革质，长椭圆形，长3～8 cm，宽1～2 cm，先端短渐尖，基部窄楔形，边缘疏生细锯齿；中脉两面凸起。总状或圆锥花序，长8～14 cm；花冠白色，卵状坛形，径约5 mm；雄蕊花丝基部被毛。蒴果球形，径3～5 mm，无毛。花期4—5月，果期7—9月。

**习性、应用及产地分布：** 生境与杜鹃相似，但更耐寒；喜半阴环境。叶有剧毒，马误食可致昏醉，故名"马醉木"。产于中国福建、浙江、江西、安徽和台湾等省，日本亦产。

**Description:** Evergreen shrubs, about 4 m tall. Bark brown, branchlets spreading, glabrous. Leaves clustered at apex of branches, leathery, oblong, 3–8 cm × 1–2 cm, apex shortly acuminate, base narrowly cuneate, margin sparsely serrate, middle vein bulging on both sides. Racemes or panicles, 8–14 cm long, corolla white, urceolate, ca. 5 mm in diam., stamens pubescent. Capsule globose, 3–5 mm in diam., glabrous. Fl. Apr–May, fr. Jul–Sep.

# 62. 紫金牛科

## 朱砂根 | *Ardisia crenata*

**别名：** 大罗伞、珍珠伞、平地木　　　　**科属：** 紫金牛科紫金牛属

**形态特征：** 常绿灌木，高 1～2 m。根断面具小红点，故称"朱砂根"；茎粗壮，少分枝。单叶互生，革质或硬纸质，有光泽，长椭圆形至椭圆状倒披针形，长 6～13 cm，宽 2～4 cm，先端急尖或渐尖，基部楔形，边缘翻卷皱波状，齿端具明显的黑色腺点；叶柄长约 1 cm。伞形或聚伞花序侧生于花枝顶端；花冠白色，略带粉红色，盛开时反卷，微香。浆果球形，径 6～7 mm，鲜红色，具腺点，经久不落。花期 5—6 月，果期 10—12 月。

**习性、应用及产地分布：** 亚热带阴性树种。喜温暖、潮湿气候，忌阳光直射；不耐寒，不耐干旱。四季常绿，秋冬红果累累，是优良的观叶、赏果植物，适宜园林假山、岩石园中配植，突显绿叶、红果之美。产于中国西藏东南部至台湾、湖北与海南等地区。

**Description:** Evergreen shrubs, 1–2 m tall. Cross section of roots with small red dots, stems stout. Leaves alternate, leathery, shiny, oblong to elliptic-oblanceolate, 6–13 cm × 2–4 cm, apex acute, base wedge-shaped, edge wrinkled, the tooth tip has distinct glandular spots; petiole 1 cm. Umbrella or cymose laterally sessile; corolla white, slightly pink, slightly fragrant. Berry globose, 6–7 mm in diam., bright red, with glandular spots. Fl. May–Jun, fr. Oct–Dec.

# 63. 柿树科

## 柿 | *Diospyros kaki*

**科属：** 柿树科柿树属

**形态特征：** 落叶乔木，高可达15 m。树皮黑灰色，方块状裂；小枝有褐黄色毛。单叶互生，阔椭圆形，长5～18 cm，宽3～9 cm，先端渐尖，全缘，革质，叶面绿色，有光泽，叶背具黄褐色柔毛，入秋部分叶片变成红色。花单生或聚生于新枝叶腋，花萼钟状，4深裂；花冠黄白色，4裂；花冠坛形或近钟形，长约2 cm，淡黄白色。果球形，径3.5～8.5 cm，成熟时橙红色，萼片肥大近木质，宿存。花期5—6月，果期9—11月。

**习性、应用及产地分布：** 温带强阳性树种。适生于温和、阳光充足之地，耐干旱，忌积水；深根性，移栽不易成活；结实早，寿命较长。树形优美，夏季叶亮绿色，入秋叶色红艳，果实满树，不易脱落，极为美观，是园林中观叶、观果树种，在公园、居民住宅区、风景区可作园路行道树、庭荫树。原产于中国长江流域，在辽宁以南各省区多有栽培。

**Description:** Deciduous trees, 15 m tall. Bark dark gray, square shape cracking, young branches yellowish-brown pubescent. Leaves alternate, broadly elliptic, 5–18 cm × 3–9 cm, apex acuminate, entire, leathery, adaxially green and glossy, abaxially covered with yellowish-brown tomentose. Flowers solitary or clustered, calyx campanulate, with 4 deeply lobed; corolla campanulate, yellowish-white, 4 lobes. Fruit globose, 3.5–8.5 cm in diam., reddish orange when mature, calyx big, persistent. Fl. May–Jun, fr. Sep–Nov.

# 君迁子 | *Diospyros lotus*

**别名：** 软枣、黑枣　　　　　　**科属：** 柿树科柿树属

**形态特征：** 落叶乔木，高可达15 m。树皮暗褐色，厚块状剥落；小枝褐色，被灰色毛，具纵裂皮孔。叶互生，椭圆形，长5～13 cm，宽2.5～6 cm，先端渐尖，叶面深绿色，幼时密生柔毛，后脱落，叶背灰色或苍白色；叶柄长7～15 mm。花小，1～3朵单生或簇生于叶腋；花萼钟形，4深裂；花冠壶形，淡黄白色。果实近球形，径1～2 cm，成熟时蓝黑色，具白蜡层，近无柄。花期5—6月，果期10—11月。

**习性、应用及产地分布：** 温带阳性树种。性强健，耐半荫，耐严寒。树干挺直，树冠圆整，浓荫覆地，秋叶变红，且适应性强，适于孤植、丛植或列植为庭荫树、园景树及"四旁"、工矿区防护林树种。产于中国东北、华北、中南及西南各地。

**Description:** Deciduous trees, 15 m tall. Bark dark brown, flaking in thick blocks; young branches brown, gray tomentose. Leaves alternate, elliptic, 5–13 cm × 2.5–6 cm, apex acuminate, adaxially dark green, densely tomentose, abaxially gray or pale; petiole 7–15 mm long. Flowers small, 1–3 flowers forming cymes, axillary, calyx, 4 lobes, corolla campanulate, light yellowish-white. Fruit nearly globose, 1–2 cm in diam., dark blue to black, epicuticular white waxy. Fl. May–Jun, fr. Oct–Nov.

# 油柿 | *Diospyros oleifera*

**科属：** 柿树科柿树属

**形态特征：** 落叶乔木，高可达15 m。树皮灰褐色，薄片状剥落，内皮白色；幼枝、叶两面及花部和果柄被灰褐色绒毛。叶互生，纸质，椭圆形至椭圆状倒卵形，长6～20 cm，宽3.5～12 cm，先端渐尖，基部圆形，边缘稍反卷，全缘，叶两面被绒毛，叶背尤密；中脉及侧脉在叶背凸起；叶柄长6～10 mm。花冠黄白色，4裂；花萼在花后增大，厚革质，径约4 cm。果扁球形或卵圆形，径约5 cm，嫩时绿色，熟时暗黄色，密被柔毛，易脱落。花期4—5月，果期8—11月。

**产地分布：** 产于中国浙江、安徽南部至两广、福建等省。

**Description:** Deciduous trees, 15 m tall. Bark grayish-brown, flaking in strips; young branches, leaves, flowers and fruit peduncle densely grey-brown tomentose. Leaves alternate, papery, elliptic to elliptically obovate, 6–20 cm × 3.5–12 cm, apex acuminate, base rounded, curving on edge, entire, adaxially and abaxially tomentose, veins abaxially raised; petiole 6–10 mm. Corolla yellowish-white, 4 lobes; calyx getting bigger after blossom, thickly leathery. Fruit globose or ovate, ca. 5 cm in diam., dark yellow when mature, densely tomentose. Fl. Apr–May, fr. Aug–Nov.

## 老鸦柿 | *Diospyros rhombifolia*

**别名：** 山柿子、野柿子

**科属：** 柿树科柿树属

**形态特征：** 落叶灌木，高2～4 m，具枝刺，幼枝淡紫色，被柔毛。叶菱形至倒卵形，长3～6 cm。花白色，单生于叶腋，花冠壶形，4裂；宿存萼片椭圆形至披针形。浆果卵球形，径约2 cm，熟时橘红色，有蜡质光泽。花期4—5月，果期9—10月。

**习性、应用及产地分布：** 喜光，较耐荫，喜湿润气候条件。花美丽芳香，果也可观，宜植于庭园观赏。产于中国浙江、江苏、安徽、江西、福建等地，生于山坡灌丛或山谷沟畔林中。

**Description:** Deciduous shrubs, 2–4 m tall, with spines. Young branches light purple, tomentose. Leaf blade rhombus to obovate, 3–6 cm long. Flowers white, solitary, axillary, corolla pot-like, 4 lobes; calyx persistent, elliptic to lanceolate. Berry ovoid, ca. 2 cm in diam., orange when mature, epicuticular waxy. Fl. Apr–May, fr. Sep–Oct.

# 64. 安息香科

## 秤锤树 | *Sinojackia xylocarpa*

**科属：** 安息香科秤锤树属

**形态特征：** 落叶小乔木，高可达7 m。小枝灰褐色，密被星状短柔毛；老枝红褐色，光滑，常纤维状脱落。叶纸质，倒卵形或椭圆形，长3～9 cm，宽2～5 cm，先端急尖，基部圆形；侧脉5～7对；叶柄长5 mm。花3～5朵成总状聚伞花序，花梗纤细，长3 cm；花萼5裂，花冠白色，裂片两面均被星状绒毛。果实卵形，红褐色，具浅棕色皮孔，顶端具圆锥状喙，长2～2.5 cm，外果皮木质，坚硬。花期3—4月，果期7—9月。

**习性、应用及产地分布：** 北亚热带树种。喜光，幼树不耐庇荫，抗寒性强，初夏白色小花随风摆动，秋季悬挂的果实宛如秤锤低垂，颇具野趣，适宜群植于山坡疏林下、林缘和庭院一隅。国家二级重点保护植物。产于中国江苏（南京）、杭州、上海、武汉等地均有栽培。

**Description:** Small trees, deciduous, 7 m tall. Young branches grayish brown, densely tomentose; old branches reddish brown, glabrous, peeling in fiber. Leaf blade papery, obovate or elliptic, 3–9 cm × 2–5 cm, apex acute, base rounded; lateral veins 5–7, petiole 5 mm long. Flowers 3–5 in cymose cluster, pedicel thin; calyx 5 lobes, corolla white, both surfaces slightly pubescent. Fruit ovoid, reddish brown, with light brown lenticels, with conical lobes on top, 2–2.5 cm, epicarp woody, hard. Fl. Mar–Apr, fr. Jul–Sep.

# 65. 木犀科

## 美国白蜡树 | *Fraxinus americana*

**别名：** 美国白蜡、美国红梣　　　　　**科属：** 木犀科白蜡树属

**形态特征：** 落叶乔木，高10～20 m。树皮微红褐色，老时纵裂；幼枝暗绿色或褐色，光滑。奇数羽状复叶具小叶5～9枚；小叶薄革质，长圆形至倒卵形，长5～15 cm，先端渐尖，基部楔形或圆形，近全缘或上部具钝齿；侧脉6～7对，中脉下凸，叶背沿中脉被毛。圆锥花序侧生于去年生枝上，长5～20 cm，先花后叶或与叶同放；无花冠。翅果长圆形，长3～5 cm，坚果圆柱形。花期4—5月，果期9—10月。

**习性、应用及产地分布：** 喜光，稍耐荫；喜温暖、湿润的气候，耐干旱瘠薄；萌芽力强，耐修剪，病虫害少；抗烟尘。可丛植或片植于庭院、街道、公园及风景区等。原产于加拿大南部和美国，中国分布几遍全国。

**Description:** Deciduous trees, 10–20 m tall. Bark reddish brown, longitudinally cracking; young branches dark green or brown, glabrous. Odd-pinnate compound leaves with 5–9 leaflets, thinly leathery, elliptic to obovate, 5–15 cm long, apex acuminate, base cuneate or rounded, margin nearly entire; lateral veins 6–7 pairs, tomentose midrib abaxially. Panicle, 5–20 cm, flower before or with leaves, without corolla. Samara elliptic, 3–5 cm, nuts cylindric. Fl. Apr–May, fr. Sep–Oct.

## 金钟花 | *Forsythia viridissima*

**别名：** 狭叶连翘　　　　　　　　　　**科属：** 木犀科连翘属

**形态特征：** 落叶灌木，高1～3 m。枝棕褐色，直立，髓薄片状。叶对生，椭圆状矩圆形，长3.5～15 cm，宽1～4 cm，先端锐尖，基部楔形，边缘中部以上具不规则锯齿；叶柄长6～12 mm；中脉及侧脉在下面凸起。花1～3朵腋生；花冠深黄色，长1.1～2.5 cm，花冠管长5～6 mm，裂片反曲，内面基部具橘黄色条纹。蒴果长1～1.5 cm，顶端喙状，2裂，宿存。花期3—4月，果期7—8月。

**习性、应用及产地分布：** 喜光，略耐荫；耐旱，不耐水湿；对土壤要求不严。早春先叶开花，满枝金黄宛如条条黄色绶带，适宜丛植、孤植于宅旁、亭阶、墙隅、篱下与路边，或溪边、池畔、岩石、假山下。产于中国江苏、安徽、浙江、江西、云南、福建等地。

**Description:** Shrubs, deciduous, 1–3 m tall. Branches brown, upright, pith lamellate. Leaves opposite, long elliptic, 3.5–15 cm × 1–4 cm, apex acute, base cuneate, margin serrated above the middle; petiole 6–12 mm; main vein and lateral veins raised abaxially. Flowers 1–3, axillary; corolla deep yellow, 1.1–2.5 cm, with orange-yellow stripes inside. Capsule 1–1.5 cm, 2 lobes, persistent. Fl. Mar–Apr, fr. Jul–Aug.

# 雪柳 | *Fontanesia fortunei*

**科属：** 木犀科雪柳属

**形态特征：** 落叶灌木至小乔木，高可达8 m。树皮灰褐色；枝灰白色，细长而直立，四棱形。单叶对生，卵状披针形，长4～12 cm，全缘，无毛。花小，花冠4裂，近达基部，白色或略带微红色，雄蕊2枚；圆锥花序顶生或腋生。小坚果扁，周围有翅。花期5—6月。

**习性、应用及产地分布：** 性强健，稍耐荫，耐寒，多生长在温暖、阴湿之地；适应性强，耐修剪。产于中国河北、陕西、山东、江苏、安徽、浙江、河南及湖北东部，在中国北方可栽作绿篱或配植于林带外缘。

**Description:** Shrubs or small trees, deciduous, 8 m tall. Barks grayish brown; branches grayish white, thin and upright, tetragonal. Leaves opposite, ovately lanceolate, 4–12 cm long, margin entire, glabrous. Flowers small, corolla 4 deep lobes, white or light reddish white, and 2 stamens; panicle axillary or subterminal. Small nut flat, with wings. Fl. May–Jun.

# 桂花 | *Osmanthus fragrans*

**别名**：木犀　　　　　　　　　　　　**科属**：木犀科木犀属

**形态特征**：常绿灌木或乔木，高3～5 m。树皮灰色，不裂。叶对生，硬革质，长椭圆形，长7～15 cm，宽2.5～4.5 cm，先端渐尖，基宽楔形，两面密布水泡状腺点，全缘或具细锯齿；侧脉6～8对；叶柄长0.8～1.5 cm。聚伞花序簇生于叶腋；花冠白色、黄色或橘黄色，极芳香，长3～4 mm；雄蕊着生于花冠筒中部。核果椭圆形，歪斜，长1～1.5 cm，熟时紫黑色。花期9—10月，果期翌年4月。

**变种、变型及栽培种**：① 四季桂，长年开花；② 银桂，花色纯白、乳白，气味较浓；③ 金桂，花色淡黄至金黄，气味浓；④ 丹桂，花色较深，橙黄至橙红色，气味适中。

**习性、应用及产地分布**：亚热带及暖温带树种。喜通风透光，稍耐荫；枝繁叶茂，终年常绿，秋季开花，香飘九里之外。中国"十大"传统名花之一。古典园林中，常与建筑物、山石相配，丛植、列植于亭台、楼阁、溪畔、水滨、桥口、山旁、亭际、墙隅、庭门、窗前附近，香随风至，尤觉宜人。原产于中国西南部，现各地广泛栽培。

**Description:** Shrubs or trees, evergreen, 3−5 m tall. Bark gray, glabrous. Leaves, opposite, hard, leathery, long-elliptic, 7−15 cm × 2.5−4.5 cm, apex acuminate, base cuneate or broadly cuneate, margin entire or usually serrulate. Cymes fascicled in leaf axils, corolla white, yellow or orange, very fragrant. Drupe purple-black, ellipsoid, oblique. Fl. Sep−Oct, fr. Apr of following year.

## 女贞 | *Ligustrum lucidum*

**科属：** 木犀科女贞属

**形态特征：** 常绿乔木，高可达 25 m。树冠倒卵形；树皮灰褐色，枝条光滑、开展。叶对生、革质、卵形或卵状椭圆形，长 7～15 cm，宽 3～7 cm，先端尖或渐尖，基部宽楔形，叶面有光泽，全缘；侧脉 4～9 对；叶柄长 1～3 cm。花密集，芳香；顶生圆锥花序大，长 10～18 cm，宽 8～25 cm，花序梗无毛，花梗极短；花冠筒长 1.5～3 mm，裂片反折，长 2～2.5 mm。果肾形，径 4～6 mm，深蓝色，熟时红黑色，被白粉。花期 6—7 月，果期 11 月至翌年 5 月。

**习性、应用及产地分布：** 亚热带树种。适应性强，喜光、耐荫、抗寒；根系发达，萌蘖力强，耐修剪、易整形。树形整齐，枝干扶疏，叶绿茂密，生长快，是园林中常用绿化树种。产于中国长江以南至华南、西南地区，朝鲜亦有分布。

**Description:** Trees, evergreen, 25 m tall. Crown obovate, bark grayish brown, branches glabrous. Leaves opposite, leathery, ovate or ovately elliptic, 7–15 cm × 3–7 cm, apex acute to acuminate, base broadly cuneate, glossy, margin entire; petiole 1–3 cm. Flowers dense, fragrant; panicles, inflorescence pedicel glabrous. Fruit reniform, deep blue-black, ripening red-black, 4–6 mm in diam. Fl. Jun–Jul, fr. Nov–May of following year.

# 金森女贞 | *Ligustrum japonicum* 'Howardii'

**科属：** 木犀科女贞属

**形态特征：** 为日本女贞系列栽培种，常绿灌木。叶对生，厚革质，椭圆形，叶缘略反卷，两面无毛；圆锥花序，花白色，花梗极短，花裂片与花冠管近等长；浆果状核果紫黑色，外被白粉。花期5—6月，果期11月。

**产地分布：** 分布于日本与中国的台湾地区。

**Description:** Shrubs, evergreen. Leaves opposite, thickly leathery, elliptic, edge revolute, glabrous on both sides; panicle, flowers white, peduncle extremely short, lobes as long as tubes; drupe, purple-black. Fl. May−Jun, fr. Nov.

## 金叶女贞 | *Ligustrum× vicaryi*

**科属：** 木犀科女贞属

**形态特征：** 半常绿灌木，高 2～3 m。是金边柳叶女贞与欧洲女贞的杂交种。叶卵状椭圆形，长 3～7 cm，嫩叶黄色，后渐变为黄绿色。顶生圆锥花序，花白色，芳香。核果成熟呈紫黑色。

**习性、应用及产地分布：** 喜光，耐修剪，抗大气污染。近年在中国北方栽培较普遍，赏其金黄色的嫩叶，但必须栽植在阳光充分处才能发挥其观叶的效果。

**Description:** Shrubs, half-evergreen, 2–3 m tall. Leaves ovately elliptic, 3–7 cm long, yellow when young, turning into yellow-green while growing. Panicle subterminal, flowers white, fragrant. Drupe, purple-black when mature.

## 红药小蜡 | *Ligustrum sinense* 'Multiflorum'

**科属：** 木犀科女贞属

**形态特征：** 半常绿灌木或小乔木，高2～4 m。小枝幼时被淡黄色短柔毛。叶纸质，椭圆形或卵状椭圆形，长3～6 cm，宽1.5～3 cm，先端短渐尖或钝而微凹，基部近圆形，叶面中脉具短柔毛；叶柄长2～8 mm。圆锥花序顶生或腋生，长4～10 cm，显具花梗，常被黄褐色短柔毛；花白色，花裂片长于花冠筒；雄蕊伸出花冠外，花药紫红色。果近球形，径5～8 mm，核果成熟呈紫黑色。花期3—6月，果期9—12月。

**习性、应用及产地分布：** 喜光，稍耐荫；生长迅速，耐修剪。枝叶细密，红药白花，十分美丽，常用于庭园作绿篱或观赏。分布于中国华中、华东和西南地区。

**Description:** Shrubs or small trees, semi-evergreen, 2–4 m tall. Young branches yellowish, pubescent. Leaves papery, elliptic or ovately elliptic, 3–6 cm × 1.5–3 cm, apex shortly acuminate or obtuse emarginate, base rounded, main veins tomentose adaxially; petiole 2–8 mm. Panicle axillary or terminal, 4–10 cm long, with obvious peduncles, yellowish brown tomentose; flowers white, lobes longer than tubes; stamens longer than corolla, anther purple-red. Fruit nearly globose, 5–8 mm in diam. Drupe, purple-black when mature. Fl. Mar–Jun, fr. Sep–Dec.

# 云南黄馨 | *Jasminum mesnyi*

**别名**：南迎春、野迎春、云南黄素馨　　**科属**：木犀科茉莉属

**形态特征**：常绿藤状灌木，高1～5 m。小枝绿色，四棱形，细长柔软，拱形下垂。叶对生，三出复叶或小枝基部单叶，纸质，长椭圆状披针形，顶端小叶较大，长2.5～6.5 cm，宽0.5～2.2 cm，先端钝或圆，具小尖头，基部楔形，叶面光滑。花常单生于叶腋，花冠漏斗状，径3～4.5 cm，黄色，裂片较花冠筒长，花冠6裂或成半重瓣。果椭圆形，径6～8 mm。花期11月至翌年4月，果期3—5月。

**习性、应用及产地分布**：喜光，稍耐荫；喜温暖、湿润气候，畏严寒。对土壤要求不严，耐干旱瘠薄；萌蘖力强。树姿婀娜，枝长柔软而垂悬，春季碧叶黄花相衬，是优良的花篱和岩石园材料。适宜植于草坪、花架、阶前、水边驳岸、台坡地、假山石旁、路边及林缘，景观效果佳。产于中国云南海拔750 m河谷、灌丛，江南各地园林常见栽培。

**Description:** Subshrubs, evergreen, 1–5 m tall. Branchlets 4-angled, green. Leaves opposite, 3-foliolate, papery, long elliptically lanceolate, terminal leaflet blade 2.5–6.5 cm × 0.5–2.2 cm, apex blunt and mucronulate, base cuneate, smooth adaxially. Flowers usually solitary, axillary, corolla funnelform, 3–4.5 cm in diam., yellow, lobes longer than tubes, corolla 6-lobed or half doubled. Berry ellipsoid, 6–8 mm in diam. Fl. Nov to Apr of following year, fr. Mar–May of following year,.

## 浓香茉莉 | *Jasminum odoratissimum*

**科属：** 木犀科茉莉属

**形态特征：** 常绿灌木，高可达8 m。枝条细长成藤本状；羽状复叶，互生，小叶5～7枚；聚伞花序，花浓香，鲜黄色。花期5—6月，果期10—11月。

**习性、应用及产地分布：** 喜光，耐半阴，抗寒耐旱，不择土壤，要求肥沃、湿润而排水好。适宜于庭院观赏、绿篱建植、高速公路两旁绿化。原产于大西洋马德拉群岛，中国华东地区时见栽培观赏。

**Description:** Shrubs, evergreen, 8 m tall. Branches thin and long; pinnate leaves, alternate, with 5–7 leaflets; cymes, flowers fragrant, bright yellow. Fl. May–Jun, fr. Oct–Nov.

## 白丁香 | *Syringa oblata* 'Alba'

**科属：** 木犀科丁香属

**形态特征：** 落叶灌木或小乔木，高可达 5 m。树皮灰褐色或灰色；小枝较粗，假二叉分枝。单叶对生，革质，广卵形，长 2～14 cm，宽 5～10 cm，叶宽常大于叶长，先端锐尖，基部近心形，全缘。圆锥花序直立，由侧芽抽生，长 6～15 cm；花冠高脚碟状或漏斗状，白色，花冠筒圆柱形，裂片直角展开；雄蕊着生于花冠筒中上部，花药内藏。蒴果长卵形，平滑，成熟时黄褐色。花期 4—5 月，果期 8—10 月。

**习性、应用及产地分布：** 温带及寒温带树种。喜光，耐半荫；耐寒性强，耐干旱瘠薄，忌酸性土；病虫害较少，可净化空气。花春季盛开，花序硕大，花色淡雅，有色有香。可植于庭院内，或孤植、丛植于路边、草坪、墙隅、林缘，或在风景区内与其他常绿树种配植。产于中国东北南部、华北、西北、山东以及四川西北部等地，长江以北各地普遍栽培。

**Description:** Shrubs or small trees, deciduous, 5 m tall. Bark grayish brown or gray; branches strong, pseudobinary branches. Leaves, opposite, leathery, broadly ovate, 2–14 cm × 5–10 cm, apex acute, base heart-shaped, margin entire. Panicle erect, 6–15 cm long, corolla funnelform, white, tube cylindric, stamens attached to middle of corolla tube, anther included. Capsule, long ovate, smooth, yellowish brown when mature. Fl. Apr–May, fr. Aug–Oct.

# 66. 马钱科

## 大叶醉鱼草 | *Buddleja davidii*

**科属**：马钱科醉鱼草属

**形态特征**：落叶灌木，高1～5 m。小枝四棱形，开展；嫩枝、叶下面、花序均密被白色星状毛。叶对生，纸质，卵状披针形至披针形，长1～20 cm，宽0.3～7.5 cm，先端渐尖，基部阔楔形至钝形，叶面深绿色，边缘疏生细锯齿；侧脉9～14对；叶柄长1～5 mm，叶柄间具2枚卵形或半圆形托叶。聚伞花序组成顶生总状圆锥花序，长4～30 cm；花冠筒细而直，长0.7～1 cm，花冠淡紫色，后变成黄白色至白色，芳香状。蒴果长椭圆形，径1.5～2 mm。花期5—10月，果期10—12月。

**习性、应用及产地分布**：温带及亚热带树种。生性强健，喜光，亦耐荫，不耐寒；根部萌芽力强。枝叶婆娑，花鲜艳密集，稍具香气，可作中型花篱植于庭园、墙隅、篱下、路旁，在空旷草地、干旱坡地丛植或作固沙树种，亦可切花观赏。分布于中国长江流域以南各省区。

**Description:** Shrubs, deciduous, 1–5 m tall. Branchlets 4-angled; young branches, abaxial leaves and inflorescence, densely scattered white tomentose. Leaves opposite, papery, ovately lanceolate to lanceolate, 1–20 cm × 0.3–7.5 cm, apex acuminate, base broad cuneate to obtuse, dark green, margin serrulate; lateral veins 9–14; petiole 1–5 mm long, with 2 ovate or half round stipules. Panicle racemous, consist of cymes, 4–30 cm long; corolla tubes thin, 0.7–1 cm long, corolla light purple or white, fragrant. Capsules, long ellipsoid, 1.5–2 mm in diam. Fl. May–Oct, fr. Oct–Dec.

## 67. 夹竹桃科

### 络石 | *Trachelospermum jasminoides*

**别名：** 万字茉莉　　　　　　　　　　**科属：** 夹竹桃科络石属

**形态特征：** 常绿攀缘藤木植物，全株具白色乳汁，藤蔓长2～10 m。枝节上常发气生根。叶革质，椭圆形至阔披针形，长2～10 cm，宽1～4.5 cm，先端锐尖或钝，基部渐狭至钝，叶面光滑，叶背被毛。花冠白色，高脚蝶形，径约2.5 cm，先端裂片偏斜呈螺旋状排列，形如风车，芳香。筒状蓇葖果细长，双生，长10～20 cm，径3～10 mm。花期6—7月，果期7—12月。

**习性、应用及产地分布：** 亚热带阳性树种。喜温暖、湿润气候，耐寒性差；对土壤要求不严，抗干旱；萌蘖性强，抗海潮风。四季常青，叶色浓绿，花繁叶茂，入秋老叶殷红。多植于枯树、假山、陡壁、支架、墙垣之旁，令其攀缘而上，自然优美；也宜作林下或常绿孤立树下的常青地被。产于中国东南部、黄河流域以南各地，朝鲜、日本亦有分布。

**Description:** Woody lianas, evergreen, with lactescent 2–10 m long. Leaves leathery, elliptic to broad lanceolate, 2–10 cm × 1–4.5 cm, apex acute or obtuse, base attenuate to obtuse, smooth adaxially, tomentose abaxially. Corolla white, 2.5 cm in diam., lobes spirally arranged, windmill-like, fragrant. Follicles linear, 10–20 cm × 3–10 mm. Fl. Jun–Jul, fr. Jul–Dec.

## 花叶蔓长春 | *Vinca major* 'Variegata'

**科属：**夹竹桃科蔓长春花属

**形态特征：**常绿蔓性灌木。单叶对生，卵形，长3～8 cm，全缘；花单生于叶腋，萼片5枚，线形；花冠蓝紫色，漏斗状，径3～5 cm，裂片5枚，开展；蓇葖果双生，直立，长约5 cm。花期5—7月。

**习性、应用及产地分布：**喜光，耐半荫，不耐寒，分蘖能力强。原产于欧洲中南部。理想的花叶兼观赏类地被材料。原产于欧洲，中国江苏、浙江和台湾等省有栽培。

**Description:** Lianas shrubs, evergreen shrubs. Leaves, opposite, ovate, 3-8 cm long, margin entire; flowers solitary, axillary, sepals 5, linear; corolla funnelform, blue-purple, 3-5 cm in diam., lobes 5; follicles double, upright, ca. 5 cm long. Fl. May-Jul.

## 夹竹桃 | *Nerium oleander*

**别名：** 红花夹竹桃、柳叶桃　　　　**科属：** 夹竹桃科夹竹桃属

**形态特征：** 常绿灌木，高可达5 m。典型三叉分枝。叶对生或3～4片轮生，狭披针形，长11～15 cm，宽2～2.5 cm，先端急尖，基部楔形，平行侧脉直达叶缘。花冠粉红色或深白色，单瓣、半重瓣或重瓣，重瓣者花冠裂片15～18枚。蓇葖果长柱形，长10～23 cm，径2～2.5 cm。花期几乎全年，夏秋最盛；果期12月至翌年1月。

**习性、应用及产地分布：** 热带及亚热带树种。喜光，耐半荫；喜温暖气候，不耐寒、耐旱。对粉尘及有毒气体有很强的吸收能力；萌蘖力强，耐修剪。碧叶如柳似竹，花色艳丽，兼有桃竹之胜，是城市绿化的极好树种，常植于庭院、公园、街头、绿地、水滨、山麓及篱下；耐烟尘、抗污染，是工矿厂区优良的防护树种。原产于伊朗、印度、尼泊尔，现广植于世界热带地区。

**Description:** Shrubs, evergreen, 5 m tall. Typical trifurcated branches. Leaves opposite or 3–4 whorled, narrowly lanceolate, 11–15 cm × 2–2.5 cm, apex acute, base cuneate, parallel lateral veins. Corolla pink or white, semi-double or double, lobes 15–18. Follicle, long cylindric, 10–23 cm long. Fl. almost all year round, fr. Dec to Jan of following year.

# 68. 茜草科

## 水团花 | *Adina pilulifera*

**别名**：水杨梅  **科属**：茜草科水团花属

**形态特征**：常绿灌木至小乔木，高可达5 m。叶对生，厚纸质，长圆状披针形，长4～12 cm，宽1.5～3 cm，先端短尖或渐尖，基部楔形，脉腋具疏毛。花小，紫红色，密集成球形头状花序，径4～6 mm，花序梗长3～4.5 cm；花萼筒被柔毛；花冠窄漏斗状，裂片卵状长圆形；子房每室具胚珠多数。果序径8～10 mm，小蒴果楔形，长2～5 mm；种子长圆形，两端具窄翅。花期6—7月，果期8—9月。

**习性、应用及产地分布**：中性树种。喜光，半阴处生长良好；好湿，不耐干旱，适生于河边、溪边和密林下。根深枝密，优良的固堤护岸树种；花尚美丽，用于庭院观赏。产于中国长江流域以南地区。

**Description:** Shrubs to small trees, evergreen, 5 m tall. Leaves, opposite, thickly papery, elliptic-lanceolate, 4–12 cm × 1.5–3 cm, apex short or acuminate, base cuneate, veins hairy. Flowers small, purplish red, dense into spherical inflorescences, 4–6 mm in diam., inflorescence pedicel 3–4.5 cm long; calyx tube tomentose, corolla narrow funnel-shaped, lobes ovate oblong. Fruit 8–10 mm in diam., capsule cuneiform, 2–5 mm long; seeds oblong, with narrow wings at both ends. Fl. Jun–Jul, fr. Aug–Sep.

# 六月雪 | *Serissa japonica*

**别名**：满天星　　　　　　　　**科属**：茜草科六月雪属

**形态特征**：半常绿矮生小灌木，高 60～90 cm；嫩枝被毛。叶对生或簇生于小枝上，革质，长椭圆形，长 6～22 mm，宽 3～6 mm，先端渐尖，基部渐狭，有光泽，全缘；两面叶脉及叶柄均被白色毛，叶柄短。花冠淡红色或白色，冠筒长约 6～12 mm，被毛，裂片较冠筒长。核果小。花期 5—7 月。

**习性、应用及产地分布**：喜温暖、荫湿环境，不耐寒；萌蘖力强，耐修剪造型。适宜在庭园路边及步道两侧植作花境、花坛、花篱，或混植于山石、岩际及林下木；亦为极好的盆景材料。产于中国长江流域以南地区至两广、香港低海拔丘陵灌丛中，日本、越南及尼泊尔亦有分布。

**Description:** Dwarf shrubs, semi-evergreen, 60–90 cm tall; young twigs hairy. Leaves opposite or fascicled on branchlets, leathery, oblong, 6–22 mm × 3–6 mm, apex acuminate, base taper, lustrous, entire. Veins and petioles tomentose, petioles short. Corolla pale red or white, corolla tube 6–12 mm long, tomentose, lobes longer than corolla tube. Drupe small. Fl. May–Jul.

# 栀子 | *Gardenia jasminoides*

**别名：** 黄栀子、山栀子　　　　　**科属：** 茜草科栀子属

**形态特征：** 常绿灌木，常丛生，高可达3 m。幼枝绿色，常被短毛。叶对生或3叶轮生，革质，全缘，倒卵状长圆形，长7～15 cm，先端渐尖或短尖，基部楔形，上面亮绿色；侧脉8～15对，在叶背明显凸起。花单生于枝顶，白色，浓香，具短梗；花萼筒具纵棱，常6裂，花萼裂片线状披针形，结果时增大；花冠高脚碟状，径3～7 cm，冠筒长3～5 cm，花裂片5～8枚，常6裂；花丝伸出；花柱长约4.5 cm。果实卵形或长椭圆形，革质或稍肉质，长1.5～7 cm，具翅状纵棱5～8条，成熟时黄色或橙红色，顶端宿存花萼长达4 cm；种子扁平。花期4—7月，果期5月至翌年2月。

**习性、应用及产地分布：** 温带及亚热带树种。喜光，亦耐荫；不耐寒；萌芽力强，耐修剪；抗大气污染。树姿端庄，枝叶茂密，花大洁白，芳香馥郁，是花叶并赏的树种。中国南方园林常作绿篱或林缘、庭石点缀，或大片群植于草坪边缘、亭阁周围及园路两侧。

**Description:** Shrubs, evergreen, up to 3 m tall. Young branches green, tomentose. Leaves opposite or in whorls of 3, leathery, entire, obovate-oblong, 7–15 cm long, apex acuminate, base cuneate, with a bright green upper surface, lateral veins obviously raised abaxially. Flowers singly born on top of branches, white, fragrant; calyx lobes linear-lanceolate, corolla tube 3–5 cm, 5–8 lobes. Fruit oval or oblong, turning yellow or orange-red when mature. Seeds flat. Fl. Apr–Jul, fr. May–Feb of following year.

## 水栀子 | *Gardenia jasminoides* var. *radicans*

**别名：** 雀舌栀子花　　　　　　　　**科属：** 茜草科栀子属

**形态特征：** 本种是栀子的变种。植株矮小，枝常平展匍地。叶较小，倒披针形，长 4～8 cm。花小，白色，重瓣，浓香。对环境没有严格要求，耐瘠薄，不耐严寒。宜植作地被材料。中国长江以南各省均有栽培。

**Description:** This species is a variant of *Gardenia jasminoides*. The plants are dwarf, branches often spread flat. Leaves, small, oblanceolate, 4–8 cm long. Flowers, white, double, fragrant.

# 69. 马鞭草科

## 马缨丹 | *Lantana camara*

**别名**：五色梅、臭草　　　　　　　　　**科属**：马鞭草科马缨丹属

**形态特征**：常绿小灌木，高1～2 m。全株具粗毛，有臭味；枝方形。单叶对生，叶卵形至卵状长圆形，长3～9 cm，宽1.5～5 cm，先端渐尖，基部圆形，两面有糙毛，上面多皱，边缘具圆钝齿。花密集成头状花序，自下而上或由外围向中心开放，花冠初开时黄色或粉红色，渐变为橘黄色或橘红色，后呈红色，故称为"五色梅"，冠筒长约1 cm。核果径约4 mm，熟时紫黑色。花期5—10月，华南地区几乎全年开花。

**习性、应用及产地分布**：亚热带至热带树种。喜温暖湿润，耐荫，不耐寒；根系发达，萌蘖力强；植株低矮，枝繁叶茂，夏季繁花锦簇，色彩柔和，美艳悦目，花期长；入秋紫果累累，花果兼赏。中国南方各地常植为花篱或配置色带图案，点缀于草坪、花坛中，在坡地、山石旁片植为地被可兼具保持水土的作用。

**Description:** Small shrubs, evergreen, 1–2 m tall. Full plant densely tomentose. Leaves opposite, ovate to ovate-oblong, 3–9 cm × 1.5–5 cm, apex acuminate, base rounded, wrinkled, margin crenate. Capitula terminal, flowers yellow or orange, often turning deep red soon after opening. Corolla tube 1 cm long. Drupe ca. 4 mm in diam., deep purple. Fl. May–Oct.

## 华紫珠 | *Callicarpa cathayana*

**科属：** 马鞭草科紫珠属

**形态特征：** 落叶灌木，高1～3 m。叶对生，长椭圆形至卵状披针形，长4～10 cm，先端渐尖，基部楔形，缘有锯齿，叶背面有红色腺点。花冠紫色，花丝与花冠近等长，花萼有星状毛，核果紫色。花期5—7月，果期8—11月。

**习性、应用及产地分布：** 亚热带树种。喜光，耐荫；喜温暖、湿润气候，较耐寒；对土壤要求不严。花果美丽，适宜植于庭园观赏。产于中国华东、中南及云南地区。

**Description:** Shrubs, deciduous, 1–3 m tall. Leaves opposite, oblong to ovate-lanceolate, 4–10 cm long, apex acuminate, base cuneate, margin serrate, leaf abaxially with red glands. Corolla purple, filaments equal longer than corolla, calyx covered with stellate tomentose. Drupe, purple. Fl. May–Jul, fr. Aug–Nov.

# 杜虹花 | *Callicarpa formosana*

**别名**：紫珠　　　　　　　　　　**科属**：马鞭草科紫珠属

**形态特征**：落叶灌木，高1～3 m。小枝、叶柄和花序均被黄色星状毛和分枝毛。叶对生，长椭圆形或卵形，长6～15 cm，宽3～8 cm，先端渐尖，基部钝圆，叶面被短硬毛，稍粗糙，叶背被灰黄色星状毛和细小黄色腺点，边缘中部以上具细齿；网脉在叶背凸起；叶柄粗壮。聚伞花序径3～4 cm；花萼杯状，被灰黄色星状毛；花冠紫色或浅紫色，长约2.5 mm，无毛；雄蕊伸出花冠外，长约为花冠3倍，子房无毛。浆果状核果小球形，径约2 mm，熟时紫色。花期5—7月，果期8—11月。

**习性、应用及产地分布**：热带阳性树种。喜高温、高湿的气候，耐干旱；适生于排水良好的湿润壤土，抗污染能力强。花朵锦簇成团，醒目壮观，秋季枝上结满紫色小果，美不胜收，适宜植于庭园或作绿篱；花可诱蝶，果可诱鸟，亦适宜植为景观生态林、水土保持林或工矿厂区绿化。产于中国浙江东南部至江西、两广、云南东南部、福建、台湾等地。

**Description:** Shrubs, deciduous, 1–3 m tall. Branchlets, petioles, and cymes densely gray-yellow stellate tomentose. Leaves opposite, oblong or ovate, 6–15 cm × 3–8 cm, apex acuminate, base obtuse, abaxially gray-yellow tomentose and yellow glandular; veins distinctly elevated abaxially. Cymes, 3–4 cm in diam., calyx cup-shaped, corolla purple to purplish, glabrous. Drupes, small globose, ca. 2 mm in diam., purple. Fl. May–Jul, fr. Aug–Nov.

# 海州常山 | *Clerodendrum trichotomum*

**别名：** 臭梧桐、后庭花　　　　　　　　**科属：** 马鞭草科赪桐属

**形态特征：** 落叶小乔木，高可达3～6 m。幼枝被黄褐色柔毛，四棱形，具棱槽；老枝灰白色，具皮孔，髓白色，有淡黄色薄片状横隔。叶对生，有臭味，阔卵形至三角状卵形，长5～16 cm，宽2～9.5 cm，先端渐尖，基部多截形，幼时两面被白色短柔毛，全缘或疏生波状齿；叶柄长3～14 cm，被短柔毛。伞房状聚伞花序，长8～18 cm；花萼紫红色，萼筒细，长约2 cm，先端5深裂；花冠白色或淡粉红色，先端5裂；雄蕊4枚，花丝与花柱同伸出花冠。核果近球形，蓝紫色，有光泽，包藏于增大的红色宿萼内。花期6—9月，果期10—11月。

**习性、应用及产地分布：** 喜光，稍耐荫；喜凉爽气候，较耐寒，耐盐碱，但不耐积水。病虫害少。丛植于庭院、路边、山坡、溪畔及石崖旁花果同赏。产于中国辽宁、甘肃、陕西以及华北、中南、西南各地，朝鲜、日本、菲律宾北部也有分布。

**Description:** Shrubs or small trees, deciduous, 3–6 m tall. Branchlets yellowish gray tomentose, 4-angled. Leaves opposite, odorous, broadly ovate to triangular-ovate, 5–16 cm × 2–9.5 cm, apex acuminate, base truncate, white pubescent, entire. Cymes, 8–18 cm long; calyx purplish red, calyx tube slender, deeply 5-lobed; corolla white or pale pink, apex 5-lobed; stamens 4, filaments and style exserted. Drupe, subglobose, blue-purple, shiny, and hidden in an enlarged red persistent calyx. Fl. Jun–Sep, fr. Oct–Nov.

# 大青 | *Clerodendrum cyrtophyllum*

**别名：** 大青叶、臭大青　　　　　　**科属：** 马鞭草科大青属

**形态特征：** 灌木或小乔木，高可达10 m。幼枝被短柔毛。单叶对生，偶有轮生，卵状椭圆形、长6～20 cm，宽3～9 cm，顶端渐尖或急尖，基部圆形或宽楔形，通常全缘，两面无毛，背面常具腺点，叶柄长2～8 cm。伞房状聚伞花序，生于枝顶或叶腋，长10～16 cm，宽20～25 cm；花小，萼杯状，花冠白色，高脚碟形或漏斗形，顶端5裂，裂片长约5 mm；雄蕊4枚，花丝与花柱伸出花冠外；子房4室，每室1个胚珠，常不完全发育。果实球形，绿色，成熟时蓝紫色，具红色宿萼。花果期6月至翌年2月。产于中国华东、中南、西南各省区。

**Description:** Shrubs or small trees, up to 10 m tall. Branchlets pubescent. Leaves opposite, occasionally whorled, ovate-elliptic, 6–20 cm × 3–9 cm, apex acuminate or acute, base rounded or broadly cuneate, margin usually entire, glabrous on both sides, abaxially glandular, petiole 2–8 cm long. Cymes, flowers small, calyx cup-shaped, corolla white, funnel-shaped, apex 5-lobed; 4 stamens and style exserted, ovary 4 rooms, 1 ovule per room, often not fully developed. Fruit spherical, green, blue-purple when mature, with red persistent calyx. Fr. Jun–Feb of following year.

# 黄荆 | *Vitex negundo*

**别名：** 五指风　　　　　　　　　**科属：** 马鞭草科牡荆属

**形态特征：** 落叶灌木或小乔木，高可达5 m。小枝四棱形，密生灰白色绒毛。掌状复叶，小叶5枚，卵状长椭圆形至披针形，中间小叶长4～13 cm，宽1～4 cm，两侧小叶依次递减，先端渐尖，基部楔形，叶面疏生短柔毛，叶背灰白色，密生细绒毛，全缘或疏生1～2粗齿。圆锥状聚伞花序顶生，长12～27 cm，花序梗密生细绒毛；花冠淡紫色，有香气，先端5裂；雄蕊与花柱均伸出花冠外。核果球形，径约5 mm，黑褐色。花期4—6月，果期7—10月。

**习性、应用及产地分布：** 喜光，耐半荫；耐寒，耐瘠薄；萌蘖力强，耐修剪。树形疏散，叶茂花繁，花色清雅，适应性强，适宜植于山坡、路旁、池边、坡地作背景材料。产于中国长江以南各省，北达秦岭、淮河。

**Description:** Shrubs or small trees, deciduous, 5 m tall. Branchlets 4-angled, densely gray-white tomentose. Leaves, palmately 5-foliolate, leaflets ovate-oblong to lanceolate, middle leaflets 4–13 cm × 1–4 cm, bilateral leaflets decreasing in size, apex acuminate, base cuneate, adaxially sparsely pubescent, abaxially grayish white, densely tomentose, margin entire or sparsely 1–2 coarsely serrate. Cymes, 12–27 cm long, peduncle densely tomentose. Corolla lilac, fragrant, apex 5-lobed; stamens and style exserted. Drupe spherical, ca. 5 mm in diam., dark brown. Fl. Apr–Jun, fr. Jul–Oct.

# 牡荆 | *Vitex negundo* var. *cannabifolia*

**科属：** 马鞭草科牡荆属

**形态特征：** 本种为黄荆的变种，形态与原种较为相似。落叶灌木；小枝四棱形，密生灰色绒毛。掌状复叶小叶5片，小叶具整齐粗锯齿，背面无毛；圆锥状聚伞花序顶生，花萼钟状，花冠淡紫色，有香气，花期4—6月。喜光，耐寒，耐瘠薄；多生于山坡、路边。产于中国华东、中南至西南各省。

**Description:** This species is a variant of *Vitex negundo*. Deciduous shrubs; branchlets 4-angled, densely gray tomentose. Leaves palmately 5-foliolate, leaflets neatly coarsely serrate, abaxially glabrous. Cymes, calyx campanulate, corolla lilac, fragrant. Fl. Apr–Jun.

# 70. 泡桐科

## 毛泡桐 | *Paulownia tomentosa*

**别名：** 紫花泡桐　　　　　　　　　　**科属：** 泡桐科泡桐属

**形态特征：** 落叶中乔木，高可达15～20 m。树冠圆伞形；树皮褐灰色，不裂；小枝皮孔明显，枝叶、花、果多被长毛。叶对生，卵形或广卵状心形，长20～30 cm，宽15～28 cm，先端渐尖或锐尖，基部心形，叶面疏被长柔毛，叶背密被具长柄的白色绒毛，全缘或波状浅裂；叶柄较长。窄圆锥形聚伞花序，长50 cm以下；花萼浅钟形，外被绒毛，深裂至中部；花冠漏斗状钟形，2唇裂，紫色或蓝紫色，内具黄色条纹和线状紫斑，外被腺毛；子房卵圆形，被腺毛。蒴果卵圆形，长3～4.5 cm。花期4—5月，果期8—9月。

**习性、应用及产地分布：** 温带强阳性树种。喜温暖气候，不耐荫庇；根系肉质，怕积水，较耐盐碱；抗大气污染，杀菌能力较强。树干通直，树冠宽广，花朵大而美丽，先叶开放，色彩绚丽，春天繁花似锦，夏日绿荫浓密，是北方平原区庭院、公园、风景区等处广泛栽植的庭荫树、行道树、园景树。产于中国黄河流域至长江流域各省，分布广泛。

**Description:** Trees, deciduous, 15–20 m tall. Bark brownish gray, not cracked; branches, leaves, flowers and fruit tomentose. Leaves opposite, ovate or broadly ovate-cordate, 20–30 cm × 15–28 cm, apex acuminate or acute, base cordate, entire, petiole longer. Cymes, 50 cm long; calyx shallowly campanulate, outside tomentose; corolla funnelform-campanulate, blue or purple, with yellow striate and purple spots. Capsule ovoid, 3–4.5 cm long. Fl. Apr–May, fr. Aug–Sep.

# 71. 紫葳科

## 梓树 | *Catalpa ovata*

**科属：** 紫葳科梓树属

**形态特征：** 落叶中乔木，高可达15～20 m。树冠伞形；树皮灰褐色，纵裂。叶广卵形或近圆形，长宽近相等，长10～25 cm，宽7～25 cm，先端渐尖，基部心形，两面粗糙，微被毛，叶背基部脉腋具紫斑，常3～5浅裂或全缘；基部掌状脉5～7条，叶脉被毛；叶柄长8～18 cm。圆锥花序，长10～20 cm；花冠长约2.5 cm，径约2 cm，淡黄色，内面具2黄色条纹及紫色细斑点；花柱丝状。蒴果线形，细长如筷，下垂，长20～30 cm，径0.5～0.7 cm，经冬不落；种子扁平。花期4—6月，果期8—11月。

**习性、应用及产地分布：** 温带阳性树种。颇耐寒，不耐干旱瘠薄，耐轻度盐碱。深根性，生长快；抗有毒气体与烟尘。树体端正，冠幅开展，叶大荫浓，春夏满树黄花，秋冬悬挂荚果细长如簪，可孤植、列植或丛植为庭荫树、行道树以及工厂绿化树种。分布极广。

**Description:** Trees, deciduous, 15−20 m tall. Crown umbelliform, bark grayish brown. Leaves, broadly ovate or subrounded, 10−25 cm × 7−25 cm, apex acuminate, base cordate, with purple spots, margin often 3−5-lobed; palmately 5−7-veined basally; petiole 8−18 cm. Panicles 10−20 cm long; corolla, pale yellow, yellow stripes and purple spots inside. Capsule, linear, slender as chopsticks, 20−30 cm long; seeds flat. Fl. Apr−Jun, fr. Aug−Nov.

# 楸树 | *Catalpa bungei*

**科属：** 紫葳科梓树属

**形态特征：** 落叶乔木，高可达 8～12 m。树皮灰褐色，树干具瘤状突起；主枝具浅细纵裂，小枝灰绿色。叶三角状卵形，长 6～15 cm，宽约 8 cm，先端尾尖，基部楔形至近圆形，两面无毛，叶背基部脉腋处具紫色腺斑，全缘或基部偶有 3～5 尖齿；叶柄长 2～8 cm。伞房状总状花序；花冠长 3～3.5 cm，径 4～5 cm，浅粉色，后呈白色，内面具 2 条黄色条纹及暗紫色斑点。蒴果线形，长 25～45 cm，径 0.5～0.6 cm。花期 5—6 月，果期 9—10 月。

**习性、应用及产地分布：** 温带及亚热带树种。喜光，幼苗耐庇荫；喜温暖、湿热气候，不耐寒，主根粗壮，根蘖和萌芽能力强。吸滞粉尘能力较强。树姿挺拔，干直荫浓，花淡紫相参差，十分美丽，宜作庭荫树及行道树。主产于中国黄河流域和长江流域。

**Description:** Trees, deciduous, 8–12 m tall. Bark grayish brown, branchlets gray-green. Leaves, triangular-ovate, 6–15 cm × 8 cm, apex long acuminate, base cuneate to subcircular, glabrous on both sides, abaxially veins with purple glandular spots, margin entire or base occasionally 3–5-lobed; petiole 2–8 cm. Corymbose; corolla, pale pink, turning white soon, yellow stripes and many purple spots at throat. Capsule, linear, 25–45 cm long. Fl. May–Jun, fr. Sep–Oct.

# 黄金树 | *Catalpa speciosa*

**别名**：白花梓树、美国楸树　　　　　**科属**：紫葳科梓树属

**形态特征**：落叶小乔木，高6～10 m。树冠伞形；树皮灰色，厚鳞片状开裂。叶宽卵形至卵状椭圆形，长15～30 cm，先端长渐尖，基部心形，叶面亮绿色，叶下密被白色短柔毛，叶背脉腋处有透明绿斑，全缘或偶有1～2浅裂；叶柄长10～15 cm。圆锥花序长约15 cm；花冠长4～5 cm，径4～6 cm，白色，内面喉部具2条黄色条纹及紫褐色细斑点。蒴果圆柱形，长30～55 cm，径1～2 cm，成熟时黑色。花期5—6月，果期8—11月。原产于美国中部及东部地区，目前全国大多数省区均有栽培。

**Description:** Small trees, deciduous, 6–10 m tall. Crown umbelliform, bark gray, cracking into deep scales. Leaves, broadly ovate to ovate-elliptic, 15–30 cm long, apex long acuminate, base cordate, abaxially veins with transparent green spots, margin entire or occasionally 1–2-lobed; petiole 10–15 cm. Panicles 15 cm long; corolla, white, yellow stripes and dark purple spots at throat. Capsule, cylindrical, 30–55 cm long, black when mature. Fl. May–Jun, fr. Aug–Nov.

## 凌霄 | *Campsis grandiflora*

**别名：** 紫葳、中国凌霄、大花凌霄　　**科属：** 紫葳科凌霄属

**形态特征：** 落叶木质藤本植物。树皮灰褐色，细条状纵裂；小枝紫褐色。奇数羽状复叶对生，具小叶 7～9 枚，卵形至卵状披针形，长 3～6 cm，宽 1.5～3 cm，先端尾状渐尖，基部阔楔形，不对称，两面无毛，边缘疏生 7～8 个粗锯齿；叶轴长 4～13 cm。聚伞状圆锥花序，长 15～20 cm；花萼长 3 cm，5 裂至中部；花冠长 6～7 cm，径 6～8 cm，外面橘黄色，内面橙红色；雄蕊着生于花冠筒近基部，花粉有毒。蒴果长如豆荚，长 10～20 cm，径约 1.5 cm；种子具透明的翅。花期 6—8 月，果期 10 月。

**习性、应用及产地分布：** 亚热带树种。喜光而稍耐荫，喜温暖、湿润气候，耐寒性较差，耐旱，忌积水；萌蘖力强。花大色艳，花期甚长，干枝虬曲多姿，为庭园中重要的棚架植物。产于中国长江流域各地，河北、山东、河南、福建、广东、广西、陕西、台湾均有栽培；日本、越南、印度、西巴基斯坦均有栽培。

**Description:** Woody vines, deciduous. Bark grayish brown; branchlets purple-brown. Leaves odd-pinnate, opposite, with leaflets 7–9, ovate to ovate-lanceolate, apex caudate-acuminate, base broadly cuneate, asymmetrical, glabrous on both sides, margin sparse 7–8 coarse serrations. Cymose-panicles, 15–20 cm long; calyx 5-lobed; corolla 6–7 cm long, orange-red. Capsule, 10–20 cm long, seeds with transparent wings. Fl. Jun–Aug, fr. Oct.

# 72. 忍冬科

## 红王子锦带 | *Weigela florida* 'Red Prince'

**科属：** 忍冬科锦带花属

**形态特征：** 落叶灌木，高可达3 m。本种是锦带花的栽培品种。树皮灰色，株型紧凑。叶椭圆形或卵状椭圆形，长5～10 cm，宽3～3.5 cm，先端渐尖，基部圆形至楔形，叶面疏被短柔毛，脉上毛较密，缘有锯齿。花萼裂至萼筒中部，裂片披针形；花冠紫红色，长3～4 cm；花丝短于花冠，花柱细长，柱头2裂。蒴果长1.5～2.5 cm，成熟时2裂，冬季宿存。花期4—6月，果期10月。

**习性、应用及产地分布：** 喜光，耐半荫，耐寒。对土壤要求不严，生长迅速，萌蘖力强；病虫害少。色泽艳丽，花期长，是优良的春夏季观花灌木，适宜植于庭园、路旁、湖畔、林缘。各地多栽培，中国华北习见。

**Description:** Deciduous shrubs, up to 3 m tall. Bark gray. Leaves, elliptic or ovate-elliptic, 5-10 cm × 3-3.5 cm, apex acuminate, base rounded to cuneate, adaxially sparsely pubescent, veins densely pubescent, margin serrate. Calyx lobes to middle of calyx tube, lobes lanceolate; corolla purplish red, 3-4 cm long; filaments shorter than corolla, style slender, 2-lobed. Capsule, 1.5-2.5 cm long, 2 cracks when mature, persistent. Fl. Apr-Jun, fr. Oct.

## 花叶锦带 | *Weigela florida* 'Variegata'

**科属：** 忍冬科锦带花属

**形态特征：** 落叶灌木。小枝沿叶柄下延具2行短柔毛；叶卵状椭圆形，脉上毛较密，缘有锯齿，叶边淡黄白色；花粉红色，柱头2裂。花期4—6月，果期10月。

**习性及产地分布：** 喜光，耐半荫，耐寒；对土壤要求不严，耐干旱瘠薄。主要分布于中国东北和华北地区。

**Description:** Shrubs, deciduous. Branchlets with 2 lines of pubescent; leaves ovate-elliptic, veins densely pubescent, margin serrate, leaf margin yellowish white; flowers pink, stigma 2-lobed. Fl. Apr-Jun, fr. Oct.

# 海仙花 | *Weigela coraeensis*

**科属：** 忍冬科锦带花属

**形态特征：** 落叶灌木。小枝粗壮，近无毛。叶广椭圆形，长8～12 cm，脉上稍被平伏毛。花冠漏斗状钟形，长2.5～4 cm，基部1/3骤狭，初开时花黄白色，后渐变成紫红色；花萼线形，裂片达基部。花无梗，数朵组成聚伞花序。蒴果2裂。花期5—6月。

**习性、应用及产地分布：** 喜光，耐荫，有一定耐寒性。原产于日本，中国华北、华东地区常见栽培。

**Description:** Shrubs, deciduous. Branchlets robust, subglabrous. Leaves, broadly oval, 8−12 cm long. Cymes, corolla funnelform-campanulate, 2.5−4 cm long, 1/3 of base abrupt narrow, flowers yellow-white, turning purplish red soon; calyx linear, lobes reaching base. Capsule, 2-lobed. Fl. May−Jun.

# 猬实 | *Kolkwitzia amabilis*

**科属：** 忍冬科猬实属

**形态特征：** 落叶灌木，高可达3 m。树皮薄片状剥落；枝梢拱曲下垂，幼枝红褐色。单叶对生，椭圆形至卵状椭圆形，长3～8 cm，宽1.5～2.5 cm，先端渐尖，基部圆形，边缘疏生浅齿或近全缘；具短柄。伞房状圆锥花序生于侧枝顶端；萼筒外密被长刚毛，先端5裂；花冠钟状，长1.5～2.5 cm，粉红色，内具黄色斑纹。2个瘦果状核果合生，有时仅1个发育，密被黄色刺刚毛，花萼宿存。花期5—6月，果期8—9月。

**习性、应用及产地分布：** 喜光，耐半荫；耐寒力强；有一定耐旱、耐瘠薄能力。花密色艳，果实形如刺猬，宜列植为花篱或孤植、丛植于草坪、角隅、路边、亭廊、假山旁及建筑附近等处。中国特有种，产于陕西、山西、甘肃、河南、湖北及安徽。

**Description:** Shrub, deciduous, 3 m tall. Branchlets reddish brown. Leaves, opposite, oblong to ovate-oblong, 3–8 cm × 1.5–2.5 cm, apex acuminate, base rounded, margin sparsely shallowly dentate or subentire. Corymbose-panicles, calyx tube with bristles abaxially, corolla campanulate, 1.5–2.5 cm long, pink, yellow-spotted. Achenes, with densely yellow thorns, hedgehog-like, calyx persistent. Fl. May–Jun, fr. Aug–Sep.

# 大花六道木 | *Abelia × grandiflora*

**科属：** 忍冬科六道木属

**形态特征：** 半常绿灌木，高可达3 m。幼枝红褐色；叶对生，卵形，表面有光泽，长2～7 cm，宽0.5～2 cm，两面疏生柔毛，全缘或中部以上具1～4对粗齿；叶柄短，基部膨大。花生于小枝顶端，花萼裂片4枚；花冠钟形，白色略带红晕，花后4枚增大花萼宿存。6月至晚秋开花不断，果期8—9月。

**习性、应用及产地分布：** 耐荫，喜凉爽气候，耐寒性强，耐干旱瘠薄；生长缓慢，根系发达，萌蘖力强，耐修剪。叶秀花美，宜配植于建筑背阴面、林下、石隙及岩石园中，或在路旁植作花篱；在山区可用作水土保持树种。产于中国东北、华北、西北等地。

**Description:** Shrubs, semi-evergreen, 3 m tall. Branchlets reddish-brown. Leaves opposite, ovate, lustrous, 2-7 cm × 0.5-2 cm, sparsely pubescent on both sides, margin entire or 1-4 pairs of coarsely serrate from middle to apex; petiole short. Flowers terminal on branchlets, calyx 4-lobed, persistent; corolla campanulate, white slightly reddish. Fl. from summer to autumn, fr. Aug-Sep.

# 金银花 | *Lonicera japonica*

**别名：** 金银藤　　　　　　　　　　　**科属：** 忍冬科忍冬属

**形态特征：** 半常绿缠绕藤木植物，茎左旋，长可达9 m。茎皮条状剥落；枝棕褐色，细长中空，幼枝暗红褐色，密被褐色粗毛及腺毛。叶对生，纸质，卵形至短圆状卵形，长3～8 cm，先端短尖或钝，基部圆形至心形，上部叶两面密被短柔毛，下部叶光滑，全缘。花双生于花梗顶端，花梗密被柔毛；花冠初开时白色，后变为黄色，二唇形，长2～6 cm，上唇4裂直立，下唇反曲；雄蕊、花柱伸出花冠外。果球形，径6～7 mm，熟时蓝黑色，有光泽。花期4—6月，秋季也常开花；果期10—11月。

**习性、应用及产地分布：** 喜光，也耐半荫；喜干燥气候，耐寒性强。对土壤要求不严，除重度盐碱土外均能生长。根系发达，萌蘖力强，茎蔓着地即能生根。植株轻盈，花朵繁密，春夏时节开花不断，色香俱备；秋末冬初果实累累，凌冬不凋。宜作篱垣、花架、栅栏等优良的垂直绿化植物，或附于山石之上，悬垂于沟边。产于中国辽宁、华北、华东及西南地区。

**Description:** Vine, semi-evergreen, 9 m long. Bark strips flaking; branches brown, slender and hollow, branchlets dark reddish brown. Leaves opposite, papery, ovate 3–8 cm, apex short tip or obtuse, base rounded or cordate, pubescent on both sides or glabrous, margin entire. Flowers, paired at the apex of peduncle, corolla white, turning yellow soon, bilabiate. Stamens and style exserted from corolla. Fruit globose, blue-black when mature, glossy. Fl. Apr–Jun, fr. Oct–Nov.

# 金银忍冬 | *Lonicera maackii*

**别名：** 金银木　　　　　　　　　　**科属：** 忍冬科忍冬属

**形态特征：** 落叶灌木或小乔木，高可达6 m。树皮灰白色，纵裂；小枝髓中空。单叶对生，纸质，卵状椭圆形至卵状披针形，长3～8 cm，先端渐尖，基部宽楔形或圆形，全缘。花双生于幼枝叶腋，花冠唇形，长1～2 cm，上唇4浅裂，下唇微翻卷，白色，后变黄色，芳香；雄蕊与花柱约为花冠的2/3。果球形，径5～6 mm，暗红色，经久不落。花期5—6月，果期8—10月。

**习性、应用及产地分布：** 性强健，喜光、耐寒；不择土壤，萌蘖性强，耐修剪，寿命长；极少有病虫害。初夏满树繁花，清雅芳香，秋季红果满枝，花果兼赏，是最常见的园林绿化树种。宜孤植或丛植于林缘、草坪、水边、建筑物周围等处。产于中国"三北"（西北、华北及东北的统称）、华东、西南地区。

**Description:** Shrubs or small trees, deciduous, up to 6 m tall. Bark grayish white, longitudinally fissured, branchlets hollow. Leaves opposite, papery, ovate-elliptic to ovate-lanceolate, 3–8 cm long, apex acuminate, base broadly cuneate or rounded, margin entire. Flowers axillary paired, corolla bilabiate, 1–2 cm long, upper 4-lobed, lower lobe slightly recurved, white and turn yellow soon, fragrant; stamens, styles about corolla 2/3 in length. Fruit globose, 5–6 mm in diam., dark red, persistent. Fl. May–Jun, fr. Aug–Oct.

# 郁香忍冬 | *Lonicera fragrantissima*

**科属：** 忍冬科忍冬属

**形态特征：** 半常绿灌木，高可达2 m。老枝灰褐色，枝具白髓心；幼枝无毛或疏被刚毛；冬芽具2枚芽鳞。叶对生，卵状椭圆形至卵状披针形，长3～8 cm，先端短尖，基部圆形或宽楔形，叶背近基部及中脉被刚毛。花成对腋生，总花梗2～10 mm，叶前开放或与叶同放，白色带粉红色，芳香；苞片条状披针形；两萼筒合生达中部以上；花冠唇形，裂片深达中部；雄蕊内藏。果球形，径约1 cm，两果合生过半，红色。花期2—4月，果期4—5月。

**习性、应用及产地分布：** 阳性树种。花期早而芳香，果色红艳，常植于庭院观赏。产于中国河北南部至安徽、浙江、湖北、江西等地。

**Description:** Shrubs, semi-evergreen, 2 m tall. Branches gray-brown, white pith; young branchlets glabrous or sparsely with bristles. Leaves opposite, ovate-elliptic to ovate-lanceolate, 3–8 cm long, apex short tip, base rounded or broadly cuneate. Flowers axillary paired, peduncle 2–10 mm. Corolla, bilabiate, white to pink, fragrant; bracteoles lanceolate; 2 calyx tubes connate to middle. Fruit globose, connate at base, red. Fl. Feb–Apr, fr. Apr–May.

# 金红久忍冬 | *Lonicera × heckrottii*

**科属：** 忍冬科忍冬属

**形态特征：** 半常绿藤本植物。老枝灰色。叶对生，卵形，表面暗绿色，背面蓝绿色；花序下的叶合成浅杯状；花冠紫红至玫瑰红色，内部金黄色，长4～5cm，二唇形，上唇4裂，下唇反卷，花10朵轮生于枝端。性强健，耐荫，耐低温。

**习性与应用：** 喜阳、耐半阴、耐旱，在肥沃、湿润的沙壤土上生长良好。花朵繁密艳丽，开花不断，是篱垣、花架、栅栏等优良的垂直绿化植物。

**Description:** Climbers, semi-evergreen. Old branches gray. Leaves opposite, ovate, adaxially dark green, abaxially blue-green, bracts and leaves under inflorescences connate into shallow cups; corolla purple to rose-red, adaxially golden yellow, 4–5 cm long, bilabiate, upper 4-lobed, lower lip recurved, flowers 10 whorls at apex of branches.

# 接骨木 | *Sambucus williamsii*

**科属：** 忍冬科接骨木属

**形态特征：** 落叶灌木或小乔木，高可达4～8 m。树皮暗灰色，密生皮孔；髓心淡黄褐色。羽状复叶对生，具小叶5～11枚，小叶卵形至长椭圆状披针形，长5～15 cm，宽1.2～7 cm，先端渐尖或尾尖，基部宽楔形，常不对称，缘有锯齿，叶揉碎后有臭味。花小而密，顶生聚伞圆锥花序，白色，与叶同放，有香气。核果球形，径约5 mm，成熟时红色，稀蓝紫色。花期5—6月，果期9—10月。

**习性、应用及产地分布：** 性强健，喜光、耐荫、耐寒、耐旱，忌水涝；生长快，根系发达，萌蘖性强，耐修剪。株形优美，枝叶繁茂，春季白花满树，夏秋红果累累，是夏季较少的观果灌木。宜植于草坪、林缘和水边，也可植为自然式绿篱。产于中国东北、华北、华东、华中、西北及其以西地区，此种分布广泛。

**Description:** Shrubs or small trees, deciduous, 4–8 m tall. Bark dark gray, with dense lenticels; pith yellowish-brown. Leaves pinnate, opposite, with leaflets 5–11, ovate to oblong-lanceolate, apex acuminate or caudate, base broadly cuneate, often asymmetrical, margin serrate, odorous. Flowers small and dense, white, cymbal-panicles, terminal, appearing simultaneously with leaves, fragrant. Drupe globose, ca. 5 mm in diam., red when mature, rarely blue-purple. Fl. May–Jun, fr. Sep–Oct.

# 日本珊瑚树 | *Viburnum odoratissimum* var. *awabuki*

**别名**：法国冬青　　　　　　　**科属**：忍冬科荚蒾属

**形态特征**：常绿中乔木，高可达10 m。枝条灰色或灰褐色，具小瘤状凸起的皮孔。叶革质，长椭圆形，长8～16 cm，宽3～6.5 cm，先端短尖至渐尖而具钝头，基部宽楔形，叶面深绿色且有光泽，中部以上疏生不规则钝齿或近全缘；羽状脉5～6对，与中脉在下面凸起。圆锥花序顶生或侧生于短枝上，长6～13.5 cm；花冠钟状，径约7 mm，白色，芳香。核果卵圆形，径5～6 mm，果红色，熟时黑色。花期5—6月，果期9—10月。

**习性、应用及产地分布**：耐荫性强；喜温暖、湿润气候，不耐寒，适应性强。适生于湿润肥沃的中性土壤。根系发达，萌蘖力强，耐修剪，易整形。枝茂叶繁，终年碧绿光亮，春季白花成簇，深秋红果状如珊瑚，花、果、叶兼赏，园林中普遍植作绿篱、绿门或绿墙。是防火隔离树带及滞尘、隔音、抗污染等多种功能的防护树种。产于中国长江流域以南。

**Description:** Trees, evergreen, up to 10 m tall. Branches gray or taupe. Leaves opposite, leathery, oblong, 8–16 cm × 3–6.5 cm, apex acuminate with blunt, base broadly cuneate, intense green adaxially, lustrous, margin subentire or sparsely irregular crenate; pinnate veins 5–6 pairs. Panicles terminal or lateral short branches, 6–13.5 cm long; corolla campanulate, ca. 7 mm in diam., white, fragrant. Drupe oval, 5–6 mm in diam., red, black when mature. Fl. May–Jun, fr. Sep–Oct.

# 木本绣球 | *Viburnum macrocephalum*

**别名**：绣球荚蒾、绣球花　　　　**科属**：忍冬科荚蒾属

**形态特征**：落叶灌木，高可达4 m。幼枝、叶柄及花序均被黄白色簇状毛，后渐变无毛。叶对生，纸质，卵形至椭圆形，长5～11 cm，宽2～5 cm，先端钝或稍尖，基部圆形或微心形，叶面初被簇状短毛，后仅中脉被毛，叶背被簇状短毛，边缘有细齿；中脉在上面略凹入，下面凸起，羽状脉5～6对，不达齿端。球形聚伞花序全由大型不孕花组成，形如绣球，径8～15 cm；花冠白色，辐射状，径1.5～4 cm，冠筒甚短，裂片5枚。花期4—5月。

**习性、应用及产地分布**：喜光，略耐荫，喜温暖、湿润气候，稍耐寒。常生于山地林间微酸性土壤，萌芽力强。树姿舒展，春日繁花如白云翻滚、积雪压枝，花落之时，白花覆地，尤饶幽趣，宜植于庭中堂前、墙下窗前、园路两侧、草坪及空旷地。中国河北、江苏、浙江、湖北、湖南、四川、福建等地均有栽培。

**Description:** Shrubs, deciduous, 4 m tall. Branchlets, petioles and inflorescences densely yellow-whitish stellate-pubescent, glabrescent. Leaves opposite, papery, ovate to elliptic, 5–11 cm × 2–5 cm, apex obtuse or slightly acute, base rounded or sparsely cordate, abaxially stellate-pubescent, margin denticulate, midvein raised abaxially, pinnate vein 5–6 pairs. Cyme, white, totally composed with the large sterile flowers, 8–15 cm in diam. Fl. Apr–May.

# 琼花 | *Viburnum macrocephalum* f. *keteleeri*

**科属**：忍冬科荚蒾属

**形态特征**：落叶灌木。本种为木本绣球的变型种。聚伞花序集生成伞房状，花序中央为两性的可育花，边缘有大型的不育花。核果椭球形，径约8 mm，先红后黑。花期4月，果期9—10月。

**产地分布**：产于中国长江中下游地区。

**Description:** Shrubs, deciduous. Cyme, the fertile bisexual flowers in the center surrounded by large sterile flowers on the edge of inflorescence. Drupe, ellipsoidal, ca. 8 mm in diam., first red then black. Fl. Apr, fr. Sep–Oct.

## 天目琼花 | *Viburnum sargentii*

**科属：** 忍冬科荚蒾属

**形态特征：** 落叶灌木，高可达3 m。树皮质厚，暗灰色，略带木栓质，浅纵裂；小枝皮孔明显。叶对生，卵圆形或宽卵形，长与宽均为6～12 cm，叶背仅脉腋处被淡黄色簇状毛及暗褐色腺点，边缘常3裂，裂片上部具粗齿，掌状三出脉；叶柄顶端具2～4个腺体。顶生复伞形式聚伞花序，周围有大型白色不孕边花，径5～10 mm；花冠乳白色，辐射状；花药紫红色。核果近球形，径8～12 mm，红色。花期5—6月，果期9—10月。

**习性、应用及产地分布：** 喜光，耐半荫，喜湿润、凉爽气候，耐寒性强。对土壤要求不严，微酸性及中性土均能生长。根系发达，移植易成活。姿态优美，边缘着生洁白的不孕花，宛若群蝶起舞，秋果似珊瑚；叶绿、花白、果红，是优良的观赏树种，适宜栽植于园林中作为点缀。产于中国东北、华北至长江流域。

**Description:** Shrubs, deciduous, 3 m tall. Bark dark gray. Leaves opposite, ovoid or broadly ovate, 6–12 cm × 6–12 cm, margin 3-lobed, with coarsely serrate from middle to apex, with 3 palmate veins; 2–4 glands on petiole. Cyme, terminal, surrounded by large white sterile flowers, 5–10 mm in diam.; corolla milky white, anther purple. Drupe, subglobose, 8–12 mm in diam., red. Fl. May–Jun, fr. Sep–Oct.

## 枇杷叶荚蒾 | *Viburnum rhytidophyllum*

**别名：**皱叶荚蒾　　　　　　　　**科属：**忍冬科荚蒾属

**形态特征：**常绿灌木或小乔木，高4～6 m。老枝黑褐色，幼枝、冬芽、叶背面、叶柄被黄褐色或红褐色厚绒毛。叶对生，革质，卵形至卵状披针形，长8～20 cm，先端钝尖，基部圆形，叶面深绿色，有光泽，全缘或具不明显小齿；叶脉深凹而呈极度皱缩状，叶背网脉凸起。聚伞花序顶生，径7～12 cm；花序轴及萼筒被黄色星状毛，花冠白色，辐射状；雄蕊5枚，伸出花冠。核果卵形，长6～8 mm，红色，成熟时黑色。花期4—5月，果期9—10月。

**习性、应用及产地分布：**喜光，耐半荫，有一定抗寒性。生长旺盛，果实美丽，宜植于园林观赏。产于中国陕西、湖北、湖南、四川和贵州等地。

**Description:** Shrubs or small trees, evergreen, 4−6 m tall. Old branches dark brown; blanchlets, winter buds, abaxially leaves and petioles covered with yellow-brownish stellate tomentose. Leaves opposite, leathery, ovate to ovate-lanceolate, 8−20 cm long, apex obtuse, base rounded, intense green adaxially, wrinkle, margin entire or serrate. Cymes terminal, 7−12 cm in diam; inflorescence axis and calyx tube yellow stellate-pubescent, corolla white. Drupe ovate, 6−8 mm, red, black when mature. Fl. Apr−May, fr. Sep−Oct.

# 地中海荚蒾 | *Viburnum tinus*

**科属：** 忍冬科荚蒾属

**形态特征：** 常绿灌木，多分枝。叶对生，卵形，全缘；聚伞花序，花蕾粉红色，于秋季现蕾，直至第2年春季开花，花冠白色；果卵形，成熟时蓝黑色。

**习性及产地分布：** 喜光，也耐荫，较耐旱。原产于欧洲地中海地区，在中国上海地区广泛种植。

**Description:** Evergreen shrubs, branchlets red. Leaves opposite, ovate, entire; cymes, flowers buds pink, corolla white; fruit ovate, blue-black when mature. Flowering from autumn to spring of following year.

## 蝴蝶绣球 | *Viburnum plicatum*

**科属：** 忍冬科荚蒾属

**形态特征：** 落叶灌木，高可达3 m。小枝密被星状柔毛。叶对生，卵圆形，边缘有锯齿，叶面羽状脉甚凹下，羽脉间又有平行小脉相连，叶背疏生星状柔毛。聚伞花序绣球形，全由大型白色不育花组成，形似蝴蝶。花期4—5月。

**习性、应用及产地分布：** 常植于庭园观赏。产于中国及日本，是中国长江流域庭园常见栽培观赏树种。

**Description:** Shrubs, deciduous, 3 m tall. Branchlets, densely stellate-pubescent. Leaves opposite, ovate, margin serrate; lateral veins pinnate, deeply impressed adaxially, veinlets parallel, sparsely stellate-pubescent abaxially. Cymes, composed of large white sterile flowers. Fl. Apr–May.

# 蝴蝶戏珠花 | *Viburnum plicatum* f. *tomentosum*

**别名**：蝴蝶树　　　　　　　　　　　**科属**：忍冬科荚蒾属

**形态特征**：落叶灌木。本种为蝴蝶绣球的变型种。叶对生，卵形，有锯齿，上面羽状脉甚凹，羽脉间有平行小脉相连；聚伞花序中部为两性花，仅边缘有大型不育花，裂片2大2小，形如蝴蝶，嬉戏珠间；秋季红色果实缀满树梢。

**产地分布**：分布于中国华东、华中、华南、西南地区。

**Description:** Shrubs, deciduous. Leaves opposite, ovate, margin serrate, lateral veins pinnate. Cymes, the fertile flowers in the center and the sterile flowers on the edge of the inflorescence. Corolla shape like the butterfly.

# 七子花 | *Heptacodium miconioides*

**科属**：忍冬科七子花属

**形态特征**：落叶灌木或小乔木，高可达7 m。茎干树皮灰白色，片状剥落。单叶对生，卵形至卵状长椭圆形，长7～16 cm，先端尾尖，基部圆形，基础3条主脉近于平行，全缘。聚伞花序对生，集成顶生的圆锥状复花序，长达15 cm；花冠白色，管状漏斗形，5深裂。核果瘦果状，具10棱。花期6—7月，果期9—11月。

**习性、应用及产地分布**：半阴性树种，喜湿润、凉爽环境，较耐寒，对土壤要求不严。是良好的观花树种。为中国特有种，产于浙江、安徽及湖北等地。

**Description:** Shrubs or small trees, deciduous, 7 m tall. Bark grayish-white, flakes off. Leaves opposite, ovate to ovate-oblong, 7–16 cm long, apex caudate, base rounded, 3 main veins nearly parallel, margin entire. Cymes opposite, terminal, 15 cm long, corolla white, tubular funnel-shaped, 5-lobed. Drupe, 10-striate. Fl. Jun–Jul, fr. Sep–Nov.

# 73. 棕榈科

## 棕榈 | *Trachycarpus fortunei*

**科属：** 棕榈科棕榈属

**形态特征：** 常绿乔木，高3～10 m。树干圆柱状，暗褐色，具圆形大叶痕。叶革质，圆扇状，掌状深裂几达基部，裂片30～50枚，线状剑形，长60～70 cm，宽2.5～4 cm，先端浅2裂，叶背面灰白色；叶柄长75～80 cm或更长，两侧具细圆齿。花雌雄异株；圆锥状肉穗花序腋生，佛焰苞棕红色，密被绒毛。雄花较小，黄绿色；雌花稍大，淡绿色，花序疏散，长80～90 cm。核果肾形，成熟时蓝黑色，被白粉，柱头宿存。花期3—5月，果期8—10月。

**习性、应用及产地分布：** 热带及亚热带树种。较耐阴，喜温暖、湿润气候；无主根，须根密集。南方园林应用广泛，可植于庭院、窗前、建筑物门前、路边及花坛之中；不宜作行道树，遮荫效果差。原产于中国，主要分布于中国长江以南各省区，日本、印度等地也有分布。

**Description:** Trees, evergreen, 3–10 m tall. Trunks cylindrical, dark brown. Leaves leathery, rounded, palmately lobes reach the base, lobes 30–50, linear sword-shaped, 60–70 cm long, apex shallow, grayish-white abaxially; petiole 75–80 cm long, margins with the fine teeth. Flowers dioecious; panicular inflorescences axillary, bracteoles red-brown, densely tomentose. Male flowers small, yellow-green; female flowers slightly larger, light green. Drupe kidney-shaped, blue-black when mature, stigma persistent. Fl. Mar–May, fr. Aug–Oct.

## 加那利海枣 | *Phoenix canariensis*

**别名：** 长叶刺葵、加拿利刺葵、加岛枣椰　　**科属：** 棕榈科刺葵属

**形态特征：** 常绿大乔木，高可达15～20 m。树干上具紧密排列的鱼鳞状叶痕。羽叶大型，长4～6 m，羽状复叶具密生羽叶100多对，裂片线状披针形，小叶基部内折，长20～40 cm，基部小叶成刺状。花小，黄褐色；肉穗花序长1 m以上。果实长椭圆形，长2～4 cm，肉薄，熟时黄色至淡红色。花期5—7月，果期8—9月。

**习性、应用及产地分布：** 热带树种。喜高温、多湿的气候和充足的阳光，耐干旱瘠薄和盐碱化，在肥沃的土壤中生长快。茎干高大雄伟，羽叶细裂而伸展，形成一密集的羽状树冠，颇显热带风光，为新优绿化树种之一，宜孤植为中心树或对植于西式建筑两侧或道路两旁。产于非洲加那利群岛及其附近地区，目前中国长江流域城市广泛种植。

**Description:** Trees, evergreen, 15–20 m tall. Leaves pinnate, 4–6 m long, 100 pairs of leaflets, 20–40 cm long, basal lobes thorny. Spadix, 1 m long; flowers small, yellowish-brown. Fruit, oblong, 2–4 cm long, yellow to pale red when mature. Fl. May–Jul, fr. Aug–Sep.

# 参 考 文 献

［1］陈有民.园林树木学［M］.北京：中国林业出版社，1990.
［2］卓丽环.园林树木学［M］.北京：中国农业出版社，2004.
［3］张天麟.园林树木1600种［M］.北京：中国建筑工业出版社，2010.
［4］申晓辉.园林树木学［M］.重庆：重庆大学出版社，2013.
［5］《中国植物志》英文修订版（Flora of China）［Z/OL］.https://www.plantplus.cn/foc，2019-11-23.

# 观赏树木中文名索引

（按汉语拼音排序）

## A

安吉拉月季 ………………… 139

## B

八角枫 ……………………… 251
八角金盘 …………………… 249
白丁香 ……………………… 276
白兰 ………………………… 75
白皮松 ……………………… 11
斑叶海桐 …………………… 121
板栗 ………………………… 49
薄叶润楠 …………………… 87
北美鹅掌楸 ………………… 80
北美枫香 …………………… 108
薜荔 ………………………… 69

## C

彩叶杞柳 …………………… 45
插田泡 ……………………… 145
茶梅 ………………………… 100
茶条槭 ……………………… 196
长叶冻绿 …………………… 217
常春油麻藤 ………………… 167
柽柳 ………………………… 231
秤锤树 ……………………… 265
池杉 ………………………… 17
重阳木 ……………………… 177
臭椿 ………………………… 184
垂柳 ………………………… 42
垂丝海棠 …………………… 136
刺槐 ………………………… 170

## D

大花六道木 ………………… 301
大花卫矛 …………………… 207
大青 ………………………… 289
大叶桉 ……………………… 234
大叶冬青 …………………… 205
大叶黄杨 …………………… 211
大叶醉鱼草 ………………… 277
地中海荚蒾 ………………… 312
棣棠 ………………………… 144
东北红豆杉 ………………… 32
杜虹花 ……………………… 287
杜鹃红山茶 ………………… 101
杜鹃花 ……………………… 256
杜梨 ………………………… 152
杜仲 ………………………… 64

## E

峨眉含笑 …………………… 78
鹅掌楸 ……………………… 79
二乔玉兰 …………………… 74
二球悬铃木 ………………… 106

## F

菲油果 ……………………… 240
榧树 ………………………… 33
粉花绣线菊 ………………… 123
粉团蔷薇 …………………… 140
风箱果 ……………………… 125
枫香 ………………………… 107
枫杨 ………………………… 38
弗吉尼亚栎 ………………… 56
扶芳藤 ……………………… 210
福建紫薇 …………………… 233
复羽叶栾树 ………………… 200

## G

柑橘 .................................. 181
宫粉梅 .............................. 149
钩栗 .................................. 50
枸骨 .................................. 203
构树 .................................. 66
光皮梾木 .......................... 246
广玉兰 .............................. 71
龟甲冬青 .......................... 206
桂花 .................................. 269

## H

海滨木槿 .......................... 223
海桐 .................................. 120
海仙花 .............................. 299
海州常山 .......................... 288
含笑 .................................. 76
豪猪刺 .............................. 93
合欢 .................................. 156
黑松 .................................. 9
红豆树 .............................. 172
红枫 .................................. 193
红花檵木 .......................... 111
红楠 .................................. 86
红千层 .............................. 235
红瑞木 .............................. 245

红王子锦带 ...................... 297
红药小蜡 .......................... 273
红叶石楠 .......................... 132
红叶腺柳 .......................... 46
红枝柴 .............................. 254
厚皮香 .............................. 102
厚叶石斑木 ...................... 153
胡颓子 .............................. 228
胡枝子 .............................. 173
蝴蝶槐 .............................. 169
蝴蝶戏珠花 ...................... 314
蝴蝶绣球 .......................... 313
花木蓝 .............................. 174
花叶锦带 .......................... 298
花叶蔓长春 ...................... 279
花叶香桃木 ...................... 238
华紫珠 .............................. 286
黄金树 .............................. 295
黄荆 .................................. 290
黄连木 .............................. 186
黄栌 .................................. 187
火棘 .................................. 127

## J

鸡爪槭 .............................. 192
檵木 .................................. 110
加那利海枣 ...................... 317

加杨 .................................. 41
夹竹桃 .............................. 280
尖叶杜英 .......................... 219
交让木 .............................. 63
接骨木 .............................. 306
结香 .................................. 227
金边胡颓子 ...................... 229
金红久忍冬 ...................... 305
金钱松 .............................. 5
金森女贞 .......................... 271
金丝梅 .............................. 104
金丝桃 .............................. 103
金塔侧柏 .......................... 21
金线柏 .............................. 27
金星球桧 .......................... 25
金叶含笑 .......................... 77
金叶女贞 .......................... 272
金叶水杉 .......................... 19
金银花 .............................. 302
金银忍冬 .......................... 303
金钟花 .............................. 267
锦鸡儿 .............................. 171
榉树 .................................. 59
君迁子 .............................. 262

## K

苦茶槭 .............................. 197

| | | |
|---|---|---|
| 苦楝 ............................... 185 | **M** | 女贞 ............................... 270 |
| 苦槠 ................................. 51 | 麻栎 ................................. 53 | |
| 阔叶十大功劳 ................. 94 | 麻叶绣线菊 ................... 122 | **P** |
| | 马甲子 ........................... 215 | 刨花楠 ............................. 88 |
| **L** | 马尾松 ............................... 8 | 枇杷 ............................... 130 |
| 蜡瓣花 ........................... 112 | 马缨丹 ........................... 285 | 枇杷叶荚蒾 ................... 311 |
| 蜡梅 ................................. 81 | 马醉木 ........................... 259 | 平枝栒子 ....................... 126 |
| 蓝冰柏 ............................. 22 | 毛白杜鹃 ....................... 257 | 葡萄 ............................... 218 |
| 榔榆 ................................. 58 | 毛白杨 ............................. 39 | |
| 老鸦柿 ........................... 264 | 毛泡桐 ........................... 292 | **Q** |
| 李 ................................... 150 | 玫瑰 ............................... 137 | 七叶树 ........................... 202 |
| 连香树 ........................... 105 | 美国白蜡树 ................... 266 | 七子花 ........................... 315 |
| 亮叶蜡梅 ......................... 82 | 美国金叶皂荚 ............... 163 | 七姊妹 ........................... 141 |
| 凌霄 ............................... 296 | 墨西哥落羽杉 ................. 16 | 榿木 ................................. 47 |
| 柳杉 ................................. 14 | 牡荆 ............................... 291 | 千金榆 ............................. 48 |
| 六月雪 ........................... 282 | 木半夏 ........................... 230 | 青冈栎 ............................. 52 |
| 龙柏 ................................. 24 | 木本绣球 ....................... 308 | 青杆 ................................... 4 |
| 龙牙花 ........................... 164 | 木芙蓉 ........................... 221 | 青檀 ................................. 62 |
| 龙爪槐 ........................... 168 | 木瓜 ............................... 135 | 琼花 ............................... 309 |
| 龙爪柳 ............................. 43 | 木槿 ............................... 222 | 楸树 ............................... 294 |
| 轮叶赤楠 ....................... 236 | 木香 ............................... 142 | |
| 罗浮槭 ........................... 199 | | **R** |
| 罗汉松 ............................. 28 | **N** | 日本花柏 ......................... 26 |
| 椤木石楠 ....................... 133 | 南方红豆杉 ..................... 31 | 日本柳杉 ......................... 13 |
| 络石 ............................... 278 | 南酸枣 ........................... 188 | 日本珊瑚树 ................... 307 |
| 落羽杉 ............................. 15 | 南天竹 ............................. 96 | 日本贴梗海棠 ............... 134 |
| | 浓香茉莉 ....................... 275 | 日本五针松 ....................... 7 |

## S

洒金东瀛珊瑚 ..... 244
三尖杉 ..... 30
三角枫 ..... 191
桑树 ..... 65
山茶 ..... 98
山胡椒 ..... 89
山麻杆 ..... 175
山梅花 ..... 115
山木香 ..... 143
山桐子 ..... 252
山皂荚 ..... 162
山楂 ..... 129
山茱萸 ..... 248
杉木 ..... 12
珊瑚朴 ..... 60
湿地松 ..... 10
石榴 ..... 242
石楠 ..... 131
柿 ..... 261
栓皮栎 ..... 54
水果蓝 ..... 255
水杉 ..... 18
水团花 ..... 281
水栀子 ..... 284
丝棉木 ..... 209
松红梅 ..... 241
溲疏 ..... 116
苏铁 ..... 2
酸橙 ..... 182
梭罗树 ..... 226

## T

台湾杉 ..... 20
桃 ..... 146
天目琼花 ..... 310
天仙果 ..... 68
天竺桂 ..... 84
甜橙 ..... 183

## W

卫矛 ..... 208
猬实 ..... 300
蚊母树 ..... 113
乌桕 ..... 176
无刺枸骨 ..... 204
无花果 ..... 67
无患子 ..... 201
梧桐 ..... 225

## X

喜树 ..... 243
细柄蕈树 ..... 109
狭叶山胡椒 ..... 90
狭叶十大功劳 ..... 95
香港四照花 ..... 247
香桃木 ..... 237
小丑火棘 ..... 128
小叶黄杨 ..... 212
小叶朴 ..... 61
小叶蚊母树 ..... 114
小叶香桃木 ..... 239
心叶椴 ..... 224
杏 ..... 148
熊掌木 ..... 250
秀丽槭 ..... 190
绣球花 ..... 118
雪柳 ..... 268
雪球冰生溲疏 ..... 117
雪松 ..... 6

## Y

羊踯躅 ..... 258
杨梅 ..... 36
野核桃 ..... 37
野花椒 ..... 178
银边八仙花 ..... 119
银荆 ..... 157
银杏 ..... 3
银芽柳 ..... 44
樱花 ..... 154
油茶 ..... 99
油柿 ..... 263

| | | |
|---|---|---|
| 榆树 ............ 57 | 皂荚 ............ 161 | 竹柏 ............ 29 |
| 羽毛枫 ............ 194 | 樟树 ............ 83 | 竹叶椒 ............ 179 |
| 羽叶槭 ............ 198 | 樟叶槭 ............ 195 | 梓树 ............ 293 |
| 玉兰 ............ 72 | 沼生栎 ............ 55 | 紫荆 ............ 159 |
| 郁李 ............ 155 | 柘树 ............ 70 | 紫穗槐 ............ 165 |
| 郁香忍冬 ............ 304 | 浙江楠 ............ 85 | 紫藤 ............ 166 |
| 元宝枫 ............ 189 | 珍珠花 ............ 124 | 紫薇 ............ 232 |
| 圆柏 ............ 23 | 栀子 ............ 283 | 紫叶碧桃 ............ 147 |
| 圆叶鼠李 ............ 216 | 枳 ............ 180 | 紫叶加拿大紫荆 ............ 160 |
| 月桂 ............ 91 | 枳椇 ............ 213 | 紫叶李 ............ 151 |
| 月季花 ............ 138 | 中华杜英 ............ 220 | 紫玉兰 ............ 73 |
| 云南黄馨 ............ 274 | 中华红叶杨 ............ 40 | 棕榈 ............ 316 |
| 云实 ............ 158 | 中华猕猴桃 ............ 97 | |
| **Z** | 舟山新木姜子 ............ 92 | |
| 枣树 ............ 214 | 朱砂根 ............ 260 | |